# Cornerstones

*Series Editor*
Steven G. Krantz, *Washington University, St. Louis, MO, USA*

Michel Willem

# Functional Analysis

## Fundamentals and Applications

Second Edition

 Birkhäuser

Michel Willem
Department of Mathematics
Université catholique de Louvain
Louvain-la-Neuve, Belgium

This book is based on the author's French edition "Principes d'Analyse Fonctionnelle", first published in France by Cassini, Paris. Copyright © 2007 by Cassini. All Rights Reserved.

ISSN 2197-182X                    ISSN 2197-1838    (electronic)
Cornerstones
ISBN 978-3-031-09151-3        ISBN 978-3-031-09149-0    (eBook)
https://doi.org/10.1007/978-3-031-09149-0

Mathematics Subject Classification: 46-XX, 46Nxx

1st edition: © Springer Science+Business Media, LLC 2013
2nd edition: © The Editor(s) (if applicable) and The Author(s), under exclusive license to Springer Nature Switzerland AG 2022

This book is published under the imprint Birkhäuser, www.birkhauser-science.com by the registered company Springer Nature Switzerland AG
The registered company address is: Gewerbestrasse 11, 6330 Cham, Switzerland

*To the memory of my father, Robert Willem,*
*and to my mother, Gilberte*
*Willem-Groeninckx*

# Preface to the Second Edition

In this second edition, some improvements have been carried out and supplementary material has been inserted.

In particular, the section on distribution theory and the chapter on "Topics in Calculus" have been completely rewritten and extended. New proofs of the density theorem in the space of functions of bounded variations and of the coarea formula are given.

In this book, the abstract integration theory depends only on Daniell's axioms and, when it is necessary, on Stone axiom, without any other assumption. In this general framework, we have added in Chap. 3 a proof of Vitali's characterization of convergence in $L^1(\Omega, \mu)$ in terms of equi-integrability and convergence in measure. In the same chapter, we have added Zabreiko's theorem on the continuity of seminorms and its applications to the closed graph theorem and to the open mapping theorem.

I want to thank my colleagues Jacques Boël, Augusto Ponce, and Jean Van Schaftingen for their suggestions, and I am particularly obliged to Cathy Brichard for her help in the realization of this second edition.

Louvain-la-Neuve, Belgium                                     Michel Willem

# Preface to the First Edition

Mathematical analysis leads to exact results by approximate computations. It is based on the notions of approximation and limit process. For instance, the derivative is the limit of differential quotients, and the integral is the limit of Riemann sums.

How to compute double limits? In some cases,

$$\int_{\Omega} \lim_{n \to \infty} u_n \, dx = \lim_{n \to \infty} \int_{\Omega} u_n \, dx,$$

$$\frac{\partial}{\partial x_k} \lim_{n \to \infty} u_n = \lim_{n \to \infty} \frac{\partial}{\partial x_k} u_n.$$

In the preceding formulas, three functional limits and one numerical limit appear. The first equality leads to the Lebesgue integral (1901), and the second to the distribution theory of Sobolev (1935) and Schwartz (1945).

In 1906, Fréchet invented an abstract framework for the limiting process: *metric spaces*. A metric space is a set $X$ with a *distance*

$$d : X \times X \to \mathbb{R} : (u, v) \mapsto d(u, v)$$

satisfying some axioms. If the real vector space $X$ is provided with a *norm*

$$X \to \mathbb{R} : u \mapsto ||u||,$$

then the formula

$$d(u, v) = ||u - v||$$

defines a distance on $X$. Finally, if the real vector space $X$ is provided with a *scalar product*

$$X \times X \to \mathbb{R} : (u, v) \mapsto (u|v),$$

then the formula

$$||u|| = \sqrt{(u|u)}$$

defines a norm on $X$.

In 1915, Fréchet defined *additive functions of sets*, or *measures*. He extended the Lebesgue integral to abstract sets. In 1918, Daniell proposed a functional definition of the abstract integral. The *elementary integral*

$$\mathcal{L} \to \mathbb{R} : u \mapsto \int_\Omega u \, d\mu,$$

defined on a vector space $\mathcal{L}$ of *elementary functions* on $\Omega$ satisfies certain axioms.

When $u$ is a nonnegative $\mu$-integrable function, its integral is given by the Cavalieri principle:

$$\int_\Omega u \, d\mu = \int_0^\infty \mu(\{x \in \Omega : u(x) > t\}) dt.$$

To measure a set is to integrate its characteristic function:

$$\mu(A) = \int_\Omega \chi_A \, d\mu.$$

In particular, the volume of a Lebesgue-measurable subset $A$ of $\mathbb{R}^N$ is defined by

$$m(A) = \int_{\mathbb{R}^N} \chi_A \, dx.$$

A *function space* is a space whose points are functions. Let $1 \leq p < \infty$. The real Lebesgue space $L^p(\Omega, \mu)$ with the norm

$$||u||_p = \left( \int_\Omega |u|^p d\mu \right)^{1/p}$$

is a *complete normed space*, or *Banach space*. The space $L^2(\Omega, \mu)$, with the scalar product

$$(u|v) = \int_\Omega uv \, d\mu,$$

is a *complete pre-Hilbert space*, or *Hilbert space*.

Duality plays a basic role in functional analysis. The *dual* of a normed space is the set of continuous linear functionals on this space. Let $1 < p < \infty$ and define $p'$, the *conjugate exponent* of $p$, by $1/p + 1/p' = 1$. The dual of $L^p(\Omega, \mu)$ is identified with $L^{p'}(\Omega, \mu)$.

*Weak derivatives* are also defined by duality. Let $f$ be a continuously differentiable function on an open subset $\Omega$ of $\mathbb{R}^N$. Multiplying $\frac{\partial f}{\partial x_k} = g$ by the *test function* $u \in \mathcal{D}(\Omega)$ and integrating by parts, we obtain

$$\int_\Omega f \frac{\partial u}{\partial x_k} dx = -\int_\Omega g\, u\, dx.$$

The preceding relation retains its meaning if $f$ and $g$ are locally integrable functions on $\Omega$. If this relation is valid for every test function $u \in \mathcal{D}(\Omega)$, then by definition, $g$ is the weak derivative of $f$ with respect to $x_k$. Like the Lebesgue integral, the weak derivatives satisfy some simple double-limit rules and are used to define some complete normed spaces, the Sobolev spaces $W^{k,p}(\Omega)$.

A *distribution* is a continuous linear functional on the space of test functions $\mathcal{D}(\Omega)$. Every locally integrable function $f$ on $\Omega$ is characterized by the distribution

$$\mathcal{D}(\Omega) \to \mathbb{R} : u \mapsto \int_\Omega fu\, dx.$$

The derivatives of the distribution $f$ are defined by

$$\langle \frac{\partial f}{\partial x_k}, u \rangle = -\langle f, \frac{\partial u}{\partial x_k} \rangle.$$

Whereas weak derivatives may not exist, distributional derivatives always exist! In this framework, Poisson's theorem in electrostatics becomes

$$-\Delta \left( \frac{1}{|x|} \right) = 4\pi \delta,$$

where $\delta$ is the Dirac measure on $\mathbb{R}^3$.

The *perimeter* of a Lebesgue-measurable subset $A$ of $\mathbb{R}^N$, defined by duality, is the *variation* of its characteristic function:

$$p(A) = \sup \left\{ \int_A \operatorname{div} v\, dx : v \in \mathcal{D}(\mathbb{R}^N; \mathbb{R}^N), \|v\|_\infty \le 1 \right\}.$$

The space of functions of *bounded variation* $BV(\mathbb{R}^N)$ contains the Sobolev space $W^{1,1}(\mathbb{R}^N)$.

Chapter 8 contains many applications to elliptic problems and to analytic or geometric inequalities. In particular, the isoperimetric inequality and the Faber–Krahn inequality are proved by purely functional-analytic methods.

The *isoperimetric inequality* in $\mathbb{R}^N$ asserts that the ball has the largest volume among all domains with fixed perimeter. In $\mathbb{R}^2$, the isoperimetric inequality is equivalent to

$$4\pi \, m(A) \le p(A)^2.$$

The *Faber–Krahn inequality* asserts that among all domains with fixed volume, the ball has the lowest fundamental eigenvalue for the Dirichlet problem. This fundamental eigenvalue is defined by

$$
\begin{aligned}
-\Delta e &= \lambda_1 e \ \text{in} \ \Omega, \\
e &> 0 \qquad \text{in} \ \Omega, \\
e &= 0 \qquad \text{on} \ \partial\Omega.
\end{aligned}
$$

Our approach is elementary and constructive. Integration theory is based on only one property: *monotone convergence*. It appears successively as an axiom, a definition, and a theorem. The inequalities of Hölder, Minkowski, and Hanner follow from the same elementary inequality, the *convexity inequality*. Weak convergence, convergence of test functions, and convergence of distributions are defined sequentially. The Hahn–Banach theorem is proved constructively in separable normed spaces and in uniformly convex smooth Banach spaces.

For the convenience of the reader, we recall the Appendix some topics in calculus. The Epilogue contains historical remarks on the close relations between functional analysis and the integral and differential calculus.

The readers must have a good knowledge of linear algebra, classical differential calculus, and the Riemann integral.

## Acknowledgments

It is a pleasure to thank Camille Debiève, Patrick Habets, Laurent Moonens, Augusto Ponce, Paolo Roselli, and Jean Van Schaftingen for their helpful comments and suggestions. It is also a pleasure to thank Sébastien de Valeriola for the beautiful drawing of the figure in Sect. 8.3. I express particularly my gratitude to Suzanne D'Addato for her outstanding mastery of TEX and her patience and to Cathy Brichard for finalizing the manuscript. Finally, I thank Ann Kostant for her exhaustive editorial work.

# Contents

# Chapter 1
# Distance

## 1.1 Real Numbers

Analysis is based on the real numbers.

**Definition 1.1.1** Let $S$ be a nonempty subset of $\mathbb{R}$. A real number $x$ is an upper bound of $S$ if for all $s \in S$, $s \leq x$. A real number $x$ is the supremum of $S$ if $x$ is an upper bound of $S$, and for every upper bound $y$ of $S$, $x \leq y$. A real number $x$ is the maximum of $S$ if $x$ is the supremum of $S$ and $x \in S$. The definitions of lower bound, infimum, and minimum are similar. We shall write $\sup S$, $\max S$, $\inf S$, and $\min S$.

Let us recall the fundamental property of $\mathbb{R}$.

**Axiom 1.1.2** Every nonempty subset of $\mathbb{R}$ that has an upper bound has a supremum.

In the extended real number system, every subset of $\mathbb{R}$ has a supremum and an infimum.

**Definition 1.1.3** The extended real number system $\overline{\mathbb{R}} = \mathbb{R} \cup \{-\infty, +\infty\}$ has the following properties:

(a) if $x \in \mathbb{R}$, then $-\infty < x < +\infty$ and $x + (+\infty) = +\infty + x = +\infty, x + (-\infty) = -\infty + x = -\infty$;
(b) if $x > 0$, then $x \cdot (+\infty) = (+\infty) \cdot x = +\infty, x \cdot (-\infty) = (-\infty) \cdot x = -\infty$;
(c) if $x < 0$, then $x \cdot (+\infty) = (+\infty) \cdot x = -\infty, x \cdot (-\infty) = (-\infty) \cdot x = +\infty$.

If $S \subset \mathbb{R}$ has no upper bound, then $\sup S = +\infty$. If $S$ has no lower bound, then $\inf S = -\infty$. Finally, $\sup \phi = -\infty$ and $\inf \phi = +\infty$.

**Definition 1.1.4** Let $X$ be a set and $F : X \to \overline{\mathbb{R}}$. We define

$$\sup_X F = \sup_{x \in X} F(x) = \sup\{F(x) : x \in X\}, \inf_X F = \inf_{x \in X} F(x) = \inf\{F(x) : x \in X\}.$$

© Springer Nature Switzerland AG 2022
M. Willem, *Functional Analysis*, Cornerstones,
https://doi.org/10.1007/978-3-031-09149-0_1

**Proposition 1.1.5** *Let X and Y be sets and $f : X \times Y \to \mathbb{R}$. Then*

$$\sup_{x \in X} \sup_{y \in Y} f(x, y) = \sup_{y \in Y} \sup_{x \in X} f(x, y), \ \sup_{x \in X} \inf_{y \in Y} f(x, y) \leq \inf_{y \in Y} \sup_{x \in X} f(x, y).$$

**Definition 1.1.6** A sequence $(x_n) \subset \mathbb{R}$ is increasing if for every $n$, $x_n \leq x_{n+1}$. The sequence $(x_n)$ is decreasing if for every $n$, $x_{n+1} \leq x_n$. The sequence $(x_n)$ is monotonic if it is increasing or decreasing.

**Definition 1.1.7** The lower limit of $(x_n) \subset \mathbb{R}$ is defined by $\varliminf_{n \to \infty} x_n = \sup_k \inf_{n \geq k} x_n$.
The upper limit of $(x_n)$ is defined by $\varlimsup_{n \to \infty} x_n = \inf_k \sup_{n \geq k} x_n$.

*Remarks*

(a) The sequence $a_k = \inf_{n \geq k} x_n$ is increasing, and the sequence $b_k = \sup_{n \geq k} x_n$ is decreasing.

(b) The lower limit and the upper limit always exist, and

$$\varliminf_{n \to \infty} x_n \leq \varlimsup_{n \to \infty} x_n.$$

**Proposition 1.1.8** *Let $(x_n), (y_n) \subset \ ]-\infty, +\infty]$ be such that $-\infty < \varliminf_{n \to \infty} x_n$ and $-\infty < \varliminf_{n \to \infty} y_n$. Then*

$$\varliminf_{n \to \infty} x_n + \varliminf_{n \to \infty} y_n \leq \varliminf_{n \to \infty} (x_n + y_n).$$

*Let $(x_n), (y_n) \subset [-\infty, +\infty[$ be such that $\varlimsup_{n \to \infty} x_n < +\infty$ and $\varlimsup_{n \to \infty} y_n < +\infty$. Then*

$$\varlimsup_{n \to \infty} (x_n + y_n) \leq \varlimsup_{n \to \infty} x_n + \varlimsup_{n \to \infty} y_n.$$

**Definition 1.1.9** A sequence $(x_n) \subset \mathbb{R}$ converges to $x \in \mathbb{R}$ if for every $\varepsilon > 0$, there exists $m \in \mathbb{N}$ such that for every $n \geq m$, $|x_n - x| \leq \varepsilon$. We then write $\lim_{n \to \infty} x_n = x$.

The sequence $(x_n)$ is a Cauchy sequence if for every $\varepsilon > 0$, there exists $m \in \mathbb{N}$ such that for every $j, k \geq m$, $|x_j - x_k| \leq \varepsilon$.

**Theorem 1.1.10** *The following properties are equivalent:*

(a) $(x_n)$ *converges;*
(b) $(x_n)$ *is a Cauchy sequence;*

*(c)* $-\infty < \varliminf_{n\to\infty} x_n \le \varlimsup_{n\to\infty} x_n < +\infty.$

If any and hence all of these properties hold, then $\lim_{n\to\infty} x_n = \varliminf_{n\to\infty} x_n = \varlimsup_{n\to\infty} x_n.$

Let us give a sufficient condition for convergence.

**Theorem 1.1.11** *Every increasing and majorized, or decreasing and minorized, sequence of real numbers converges.*

*Remark* Every increasing sequence of real numbers that is not majorized converges in $\overline{\mathbb{R}}$ to $+\infty$. Every decreasing sequence of real numbers that is not minorized converges in $\overline{\mathbb{R}}$ to $-\infty$. Hence, if $(x_n)$ is increasing, then

$$\lim_{n\to\infty} x_n = \sup_n x_n,$$

and if $(x_n)$ is decreasing, then

$$\lim_{n\to\infty} x_n = \inf_n x_n.$$

In particular, for every sequence $(x_n) \subset \overline{\mathbb{R}}$,

$$\varliminf_{n\to\infty} x_n = \lim_{k\to\infty} \inf_{n\ge k} x_n$$

and

$$\varlimsup_{n\to\infty} x_n = \lim_{k\to\infty} \sup_{n\ge k} x_n.$$

**Definition 1.1.12** The series $\displaystyle\sum_{n=0}^{\infty} x_n$ converges, and its sum is $x \in \mathbb{R}$ if the sequence $\displaystyle\sum_{n=0}^{k} x_n$ converges to $x$. We then write $\displaystyle\sum_{n=0}^{\infty} x_n = x.$

**Theorem 1.1.13** *The following statements are equivalent:*

*(a)* $\displaystyle\sum_{n=0}^{\infty} x_n$ *converges;*

*(b)* $\displaystyle\lim_{\substack{j\to\infty \\ j<k}} \sum_{n=j+1}^{k} x_n = 0.$

**Theorem 1.1.14** *Let $(x_n)$ be such that $\displaystyle\sum_{n=0}^{\infty}|x_n|$ converges. Then $\displaystyle\sum_{n=0}^{\infty}x_n$ converges and*

$$\left|\sum_{n=0}^{\infty}x_n\right| \leq \sum_{n=0}^{\infty}|x_n|.$$

## 1.2   Metric Spaces

Metric spaces were created by Maurice Fréchet in 1906.

**Definition 1.2.1**  A distance on a set $X$ is a function

$$X \times X \to \mathbb{R} : (u, v) \to d(u, v)$$

such that

$(\mathcal{D}_1)$    for every $u, v \in X, d(u, v) = 0 \Longleftrightarrow u = v$;
$(\mathcal{D}_2)$    for every $u, v \in X, d(u, v) = d(v, u)$;
$(\mathcal{D}_3)$    (triangle inequality) for every $u, v, w \in X, d(u, w) \leq d(u, v) + d(v, w)$.

A metric space is a set together with a distance on that set.

*Examples*

1. Let $(X, d)$ be a metric space and let $S \subset X$. The set $S$ together with $d$ (restricted to $S \times S$) is a metric space.
2. Let $(X_1, d_1)$ and $(X_2, d_2)$ be metric spaces. The set $X_1 \times X_2$ together with

$$d((x_1, x_2), (y_1, y_2)) = \max\{d_1(x_1, y_1), d_2(x_2, y_2)\}$$

   is a metric space.
3. We define the distance on the space $\mathbb{R}^N$ to be

$$d(x, y) = \max\{|x_1 - y_1|, \ldots, |x_n - y_n|\}.$$

4. We define the distance on the space $C([0, 1]) = \{u : [0, 1] \to \mathbb{R} : u$ is continuous$\}$ to be

$$d(u, v) = \max_{x \in [0,1]} |u(x) - v(x)|.$$

**Definition 1.2.2** Let $X$ be a metric space. A sequence $(u_n) \subset X$ converges to $u \in X$ if

$$\lim_{n \to \infty} d(u_n, u) = 0.$$

We then write $\lim_{n \to \infty} u_n = u$ or $u_n \to u, n \to \infty$. The sequence $(u_n)$ is a Cauchy sequence if

$$\lim_{j,k \to \infty} d(u_j, u_k) = 0.$$

The sequence $(u_n)$ is bounded if

$$\sup_n d(u_0, u_n) < \infty.$$

**Proposition 1.2.3** *Every convergent sequence is a Cauchy sequence. Every Cauchy sequence is a bounded sequence.*

**Proof** If $(u_n)$ converges to $u$, then by the triangle inequality, it follows that

$$0 \le d(u_j, u_k) \le d(u_j, u) + d(u, u_k)$$

and $\lim_{j,k \to \infty} d(u_j, u_k) = 0$.

If $(u_n)$ is a Cauchy sequence, then there exists $m$ such that for $j, k \ge m$, $d(u_j, u_k) \le 1$. We obtain for every $n$ that

$$d(u_0, u_n) \le \max\{d(u_0, u_1), \ldots, d(u_0, u_{m-1}), d(u_0, u_m) + 1\}. \qquad \square$$

**Definition 1.2.4** A sequence $(u_{n_j})$ is a subsequence of a sequence $(u_n)$ if for every $j, n_j < n_{j+1}$.

**Definition 1.2.5** Let $X$ be a metric space. The space $X$ is complete if every Cauchy sequence in $X$ converges. The space $X$ is precompact if every sequence in $X$ contains a Cauchy subsequence. The space $X$ is compact if every sequence in $X$ contains a convergent subsequence.

*Remark*

(a) Completeness allows us to prove the convergence of a sequence without using the limit.
(b) Compactness will be used to prove existence theorems and to find hidden uniformities.

The proofs of the next propositions are left to the reader.

**Proposition 1.2.6** *Every Cauchy sequence containing a convergent subsequence converges. Every subsequence of a convergent, Cauchy, or bounded sequence satisfies the same property.*

**Proposition 1.2.7** *A metric space is compact if and only if it is precompact and complete.*

**Theorem 1.2.8** *The real line $\mathbb{R}$, with the usual distance, is complete.*

*Example (A Noncomplete Metric Space)* We define the distance on $X = C([0, 1])$ to be

$$d(u, v) = \int_0^1 |u(x) - v(x)| \, dx.$$

Every sequence $(u_n) \subset X$ such that

(a)  for every $x$ and for every $n$, $u_n(x) \le u_{n+1}(x)$;

(b)  $\sup_n \int_0^1 u_n(x)dx = \lim_{n \to \infty} \int_0^1 u_n(x)dx < +\infty$;

is a Cauchy sequence. Indeed, we have that

$$\lim_{j,k \to \infty} \int_0^1 |u_j(x) - u_k(x)|dx = \lim_{j,k \to \infty} |\int_0^1 (u_j(x) - u_k(x))dx| = 0.$$

But $X$ with $d$ is not complete, since the sequence defined by

$$u_n(x) = \min\{n, 1/\sqrt{x}\}$$

satisfies (a) and (b) but is not convergent. Indeed, assuming that $(u_n)$ converges to $u$ in $X$, we obtain, for $0 < \varepsilon < 1$, that

$$\int_\varepsilon^1 |u(x) - 1/\sqrt{x}|dx = \lim_{n \to \infty} \int_\varepsilon^1 |u(x) - u_n(x)|dx \le \lim_{n \to \infty} \int_0^1 |u(x) - u_n(x)|dx = 0.$$

But this is impossible, since $u(x) = 1/\sqrt{x}$ has no continuous extension at 0.

**Definition 1.2.9** Let $X$ be a metric space, $u \in X$, and $r > 0$. The open and closed balls of center $u$ and radius $r$ are defined by

$$B(u, r) = \{v \in X : d(v, u) < r\}, \quad B[u, r] = \{v \in X : d(v, u) \le r\}.$$

The subset $S$ of $X$ is open if for all $u \in S$, there exists $r > 0$ such that $B(u, r) \subset S$. The subset $S$ of $X$ is closed if $X \setminus S$ is open.

*Example* Open balls are open; closed balls are closed.

**Proposition 1.2.10** *The union of every family of open sets is open. The intersection of a finite number of open sets is open. The intersection of every family of closed sets is closed. The union of a finite number of closed sets is closed.*

**Proof** The properties of open sets follow from the definition. The properties of closed sets follow by considering complements.    □

**Definition 1.2.11** Let $S$ be a subset of a metric space $X$. The interior of $S$, denoted by $\overset{\circ}{S}$, is the largest open set of $X$ contained in $S$. The closure of $S$, denoted by $\overline{S}$, is the smallest closed set of $X$ containing $S$. The boundary of $S$ is defined by $\partial S = \overline{S} \setminus \overset{\circ}{S}$. The set $S$ is dense if $\overline{S} = X$.

**Proposition 1.2.12** *Let $X$ be a metric space, $S \subset X$, and $u \in X$. Then the following properties are equivalent:*

*(a)* $u \in \overline{S}$;
*(b) for all $r > 0$, $B(u, r) \cap S \neq \phi$;*
*(c) there exists $(u_n) \subset S$ such that $u_n \to u$.*

**Proof** It is clear that (b) ⇔ (c). Assume that $u \notin \overline{S}$. Then there exists a closed subset $F$ of $X$ such that $u \notin F$ and $S \subset F$. By definition, then exists $r > 0$ such that $B(u, r) \cap S = \phi$. Hence (b) implies (a). If there exists $r > 0$ such that $B(u, r) \cap S = \phi$, then $F = X \setminus B(u, r)$ is a closed subset containing $S$. We conclude that $u \notin \overline{S}$. Hence (a) implies (b).    □

**Theorem 1.2.13 (Baire's Theorem)** *In a complete metric space, every intersection of a sequence of open dense subsets is dense.*

**Proof** Let $(U_n)$ be a sequence of dense open subsets of a complete metric space $X$. We must prove that for every open ball $B$ of $X$, $B \cap \left( \cap_{n=0}^{\infty} U_n \right) \neq \phi$. Since $B \cap U_0$ is open (Proposition 1.2.10) and nonempty (density of $U_0$), there is a closed ball $B[u_0, r_0] \subset B \cap U_0$. By induction, for every $n$, there is a closed ball

$$B[u_n, r_n] \subset B(u_{n-1}, r_{n-1}) \cap U_n$$

such that $r_n \leq 1/n$. Then $(u_n)$ is a Cauchy sequence. Indeed, for $j, k \geq n$, $d(u_j, u_k) \leq 2/n$. Since $X$ is complete, $(u_n)$ converges to $u \in X$. For $j \geq n$, $u_j \in B[u_n, r_n]$, so that for every $n$, $u \in B[u_n, r_n]$. It follows that $u \in B \cap (\cap_{n=0}^{\infty} U_n)$.    □

*Example* Let us prove that $\mathbb{R}$ is uncountable. Assume that $(r_n)$ is an enumeration of $\mathbb{R}$. Then for every $n$, the set $U_n = \mathbb{R} \setminus \{r_n\}$ is open and dense. But then $\cap_{n=1}^{\infty} U_n$ is dense and empty. This is a contradiction.

**Definition 1.2.14** Let $X$ be a metric space with distance $d$ and let $S \subset X$. The subset $S$ is complete, precompact, or compact if $S$ with distance $d$ is complete, precompact, or compact. A covering of $S$ is a family $\mathcal{F}$ of subsets of $X$ such that the union of $\mathcal{F}$ contains $S$.

**Proposition 1.2.15** *Let $X$ be a complete metric space and let $S \subset X$. Then $S$ is closed if and only if $S$ is complete.*

**Proof** It suffices to use Proposition 1.2.12 and the preceding definition.                    □

**Theorem 1.2.16 (Fréchet's Criterion, 1910)** *Let $X$ be a metric space and let $S \subset X$. The following properties are equivalent:*

*(a)  $S$ is precompact;*
*(b)  for every $\varepsilon > 0$, there is a finite covering of $S$ by balls of radius $\varepsilon$.*

**Proof** Assume that $S$ satisfies (b). We must prove that every sequence $(u_n) \subset S$ contains a Cauchy subsequence. Cantor's diagonal argument will be used. There is a ball $B_1$ of radius 1 containing a subsequence $(u_{1,n})$ from $(u_n)$. By induction, for every $k$, there is a ball $B_k$ of radius $1/k$ containing a subsequence $(u_{k,n})$ from $(u_{k-1,n})$. The sequence $v_n = u_{n,n}$ is a Cauchy sequence. Indeed, for $m, n \geq k$, $v_m, v_n \in B_k$ and $d(v_m, v_n) \leq 2/k$.

Assume that (b) is not satisfied. There then exists $\varepsilon > 0$ such that $S$ has no finite covering by balls of radius $\varepsilon$. Let $u_0 \in S$. There is $u_1 \in S \setminus B[u_0, \varepsilon]$. By induction, for every $k$, there is

$$u_k \in S \setminus \bigcup_{j=0}^{k-1} B[u_j, \varepsilon].$$

Hence for $j < k$, $d(u_j, u_k) \geq \varepsilon$, and the sequence $(u_n)$ contains no Cauchy subsequence.                    □

Every precompact space is *separable*.

**Definition 1.2.17** A metric space is separable if it contains a countable dense subset.

**Proposition 1.2.18** *Let $X$ and $Y$ be separable metric spaces, and let $S$ be a subset of $X$.*

*(a)  The space $X \times Y$ is separable.*
*(b)  The space $S$ is separable.*

**Proof** Let $(e_n)$ and $(f_n)$ be sequences dense in $X$ and $Y$. The family $\{(e_n, f_k) : (n, k) \in \mathbb{N}^2\}$ is countable and dense in $X \times Y$. Let

$$\mathcal{F} = \{(n, k) \in \mathbb{N}^2 : k \geq 1, B(e_n, 1/k) \cap S \neq \phi\}.$$

For every $(n, k) \in \mathcal{F}$, we choose $f_{n,k} \in B(e_n, 1/k) \cap S$. The family $\{f_{n,k} : (n, k) \in \mathcal{F}\}$ is countable and dense in $S$. □

## 1.3   Continuity

Let us define continuity using distances.

**Definition 1.3.1** Let $X$ and $Y$ be metric spaces. A mapping $u : X \to Y$ is continuous at $y \in X$ if for every $\varepsilon > 0$, there exists $\delta > 0$ such that

$$\sup\{d_Y(u(x), u(y)) : x \in X, d_X(x, y) \leq \delta\} \leq \varepsilon. \tag{*}$$

The mapping $u$ is continuous if it is continuous at every point of $X$. The mapping $u$ is uniformly continuous if for every $\varepsilon > 0$, there exists $\delta > 0$ such that

$$\omega_u(\delta) = \sup\{d_Y(u(x), u(y)) : x, y \in X, d_X(x, y) \leq \delta\} \leq \varepsilon.$$

The function $\omega_u$ is the modulus of continuity of $u$.

*Remark* It is clear that uniform continuity implies continuity. In general, the converse is false. We shall prove the converse when the domain of the mapping is a compact space.

*Example* The distance $d : X \times X \to \mathbb{R}$ is uniformly continuous, since

$$|d(x_1, x_2) - d(y_1, y_2)| \leq 2 \max\{d(x_1, y_1), d(x_2, y_2)\}.$$

**Lemma 1.3.2** *Let $X$ and $Y$ be metric spaces, $u : X \to Y$, and $y \in X$. The following properties are equivalent:*

*(a) $u$ is continuous at $y$;*
*(b) if $(y_n)$ converges to $y$ in $X$, then $(u(y_n))$ converges to $u(y)$ in $Y$.*

**Proof** Assume that $u$ is not continuous at $y$. Then there is $\varepsilon > 0$ such that for every $n$, there exists $y_n \in X$ such that

$$d_X(y_n, y) \leq 1/n \quad \text{and} \quad d_Y(u(y_n), u(y)) > \varepsilon.$$

But then $(y_n)$ converges to $y$ in $X$ and $(u(y_n))$ is not convergent to $u(y)$.

Let $u$ be continuous at $y$ and $(y_n)$ converging to $y$. Let $\varepsilon > 0$. There exists $\delta > 0$ such that (*) is satisfied, and there exists $m$ such that for every $n \geq m$, $d_X(y_n, y) \leq \delta$. Hence for $n \geq m$, $d_Y(u(y_n), u(y)) \leq \varepsilon$. Since $\varepsilon > 0$ is arbitrary, $(u(y_n))$ converges to $u(y)$. □

**Proposition 1.3.3** *Let X and Y be metric spaces, K a compact subset of X, and u :
X → Y a continuous mapping, constant on X \ K. Then u is uniformly continuous.*

**Proof** Assume that $u$ is not uniformly continuous. Then there is $\varepsilon > 0$ such that for
every $n$, there exist $x_n \in X$ and $y_n \in K$ such that

$$d_X(x_n, y_n) \le 1/n \text{ and } d_Y(u(x_n), u(y_n)) > \varepsilon.$$

By compactness, there is a subsequence $(y_{n_k})$ converging to $y$. Hence $(x_{n_k})$
converges also to $y$. It follows from the continuity of $u$ at $y$ and from the preceding
lemma that

$$\varepsilon \le \overline{\lim_{k \to \infty}} \, d_Y(u(x_{n_k}), u(y_{n_k}))$$

$$\le \lim_{k \to \infty} d_Y(u(x_{n_k}), u(y)) + \lim_{k \to \infty} d_Y(u(y), u(y_{n_k})) = 0.$$

This is a contradiction.                                                        □

**Lemma 1.3.4** *Let X be a set and F : X → ]−∞, +∞] a function. Then there
exists a sequence $(y_n) \subset X$ such that $\lim_{n \to \infty} F(y_n) = \inf_X F$. The sequence $(y_n)$ is
called a minimizing sequence.*

**Proof** If $c = \inf_X F \in \mathbb{R}$, then for every $n \ge 1$, there exists $y_n \in X$ such that

$$c \le F(y_n) \le c + 1/n.$$

If $c = -\infty$, then for every $n \ge 1$, there exists $y_n \in X$ such that

$$F(y_n) \le -n.$$

In both cases, the sequence $(y_n)$ is a minimizing sequence. If $c = +\infty$, the result is
obvious.                                                                       □

**Proposition 1.3.5** *Let X be a compact metric space, and let F : X → ℝ be a
continuous function. Then F is bounded, and there exists $y, z \in X$ such that*

$$F(y) = \min_X F, \quad F(z) = \max_X F.$$

**Proof** Let $(y_n) \subset X$ be a minimizing sequence: $\lim_{n \to \infty} F(y_n) = \inf_X F$. There is a
subsequence $(y_{n_k})$ converging to $y$. We obtain

$$F(y) = \lim_{k \to \infty} F(y_{n_k}) = \inf_X F.$$

Hence $y$ minimizes $F$ on $X$. To prove the existence of $z$, consider $-F$.            □

The preceding proof suggests a generalization of continuity.

**Definition 1.3.6** Let $X$ be a metric space. A function $F : X \to \ ]-\infty, +\infty]$ is lower semicontinuous (l.s.c.) at $y \in X$ if for every sequence $(y_n)$ converging to $y$ in $X$,

$$F(y) \le \varliminf_{n \to \infty} F(y_n).$$

The function $F$ is lower semicontinuous if it is lower semicontinuous at every point of $X$. A function $F : X \to [-\infty, +\infty[$ is upper semicontinuous (u.s.c.) at $y \in X$ if for every sequence $(y_n)$ converging to $y$ in $X$,

$$\varlimsup_{n \to \infty} F(y_n) \le F(y).$$

The function $F$ is upper semicontinuous if it is upper semicontinuous at every point of $X$.

*Remark* A function $F : X \to \mathbb{R}$ is continuous at $y \in X$ if and only if $F$ is both l.s.c. and u.s.c. at $y$.

Let us generalize the preceding proposition.

**Proposition 1.3.7** *Let $X$ be a compact metric space and let $F : X \to \ ]-\infty, \infty]$ be an l.s.c. function. Then $F$ is bounded from below, and there exists $y \in X$ such that*

$$F(y) = \min_{X} F.$$

*Proof* Let $(y_n) \subset X$ be a minimizing sequence. There is a subsequence $(y_{n_k})$ converging to $y$. We obtain

$$F(y) \le \varliminf_{k \to \infty} F(y_{n_k}) = \inf_{X} F.$$

Hence $y$ minimizes $F$ on $X$.                                                         □

When $X$ is not compact, the situation is more delicate.

**Theorem 1.3.8 (Ekeland's Variational Principle)** *Let $X$ be a complete metric space, and let $F : X \to \ ]-\infty, +\infty]$ be an l.s.c. function such that $c = \inf_X F \in \mathbb{R}$. Assume that $\varepsilon > 0$ and $z \in X$ are such that*

$$F(z) \le \inf_{X} F + \varepsilon.$$

*Then there exists $y \in X$ such that*

*(a)  $F(y) \le F(z)$;*
*(b)  $d(y, z) \le 1$;*

*(c) for every $x \in X \setminus \{y\}$, $F(y) - \varepsilon\, d(x, y) < F(x)$.*

**Proof** Let us define inductively a sequence $(y_n)$. We choose $y_0 = z$ and

$$y_{n+1} \in S_n = \{x \in X : F(x) \leq F(y_n) - \varepsilon\, d(y_n, x)\}$$

such that

$$F(y_{n+1}) - \inf_{S_n} F \leq \frac{1}{2}\left[F(y_n) - \inf_{S_n} F\right]. \tag{$*$}$$

Since for every $n$,

$$\varepsilon\, d(y_n, y_{n+1}) \leq F(y_n) - F(y_{n+1}),$$

we obtain

$$c \leq F(y_{n+1}) \leq F(y_n) \leq F(y_0) = F(z),$$

and for every $k \geq n$,

$$\varepsilon\, d(y_n, y_k) \leq F(y_n) - F(y_k). \tag{$**$}$$

Hence

$$\lim_{\substack{n \to \infty \\ k \geq n}} d(y_n, y_k) = 0.$$

Since $X$ is complete, the sequence $(y_n)$ converges to $y \in X$. Since $F$ is l.s.c., we have

$$F(y) \leq \lim_{n \to \infty} F(y_n) \leq F(z).$$

It follows from $(**)$ that for every $n$,

$$\varepsilon\, d(y_n, y) \leq F(y_n) - F(y).$$

In particular, for every $n$, $y \in S_n$, and for $n = 0$,

$$\varepsilon\, d(z, y) \leq F(z) - F(y) \leq c + \varepsilon - c = \varepsilon.$$

Finally, assume that

$$F(x) \leq F(y) - \varepsilon\, d(x, y).$$

The fact that $y \in S_n$ implies that $x \in S_n$. By $(*)$, we have

$$2F(y_{n+1}) - F(y_n) \le \inf_{S_n} F \le F(x),$$

so that

$$F(y) \le \lim_{n \to \infty} F(y_n) \le F(x).$$

We conclude that $x = y$, because

$$\varepsilon \, d(x, y) \le F(y) - F(x) \le 0. \qquad \square$$

**Definition 1.3.9** Let $X$ be a set. The upper envelope of a family of functions $F_j :$ $X \to \, ]-\infty, \infty]$, $j \in J$, is defined by

$$\left( \sup_{j \in J} F_j \right)(x) = \sup_{j \in J} F_j(x).$$

**Proposition 1.3.10** *The upper envelope of a family of l.s.c. functions at a point of a metric space is l.s.c. at that point.*

**Proof** Let $F_j : X \to \, ]-\infty, +\infty]$ be a family of l.s.c. functions at $y$. By Proposition 1.1.5, we have, for every sequence $(y_n)$ converging to $y$,

$$\sup_j F_j(y) \le \sup_j \varliminf_{n \to \infty} F_j(y_n) = \sup_j \sup_k \inf_m F_j(y_{m+k})$$

$$\le \sup_k \inf_m \sup_j F_j(y_{m+k}) = \varliminf_{n \to \infty} \sup_j F_j(y_n).$$

Hence $\sup_j F_j$ is l.s.c. at $y$. $\qquad \square$

**Proposition 1.3.11** *The sum of two l.s.c. functions at a point of a metric space is l.s.c. at this point.*

**Proof** Let $F, G : X \to \, ]-\infty, \infty]$ be l.s.c. at $y$. By Proposition 1.1.8, we have for every sequence $(y_n)$ converging to $y$ that

$$F(y) + G(y) \le \varliminf_{n \to \infty} F(y_n) + \varliminf_{n \to \infty} G(y_n) \le \varliminf_{n \to \infty} (F(y_n) + G(y_n)).$$

Hence $F + G$ is l.s.c. at $y$. $\qquad \square$

**Proposition 1.3.12** *Let* $F : X \to ]-\infty, \infty]$. *The following properties are equivalent:*

*(a)  F is l.s.c;*
*(b)  for every* $t \in \mathbb{R}$, $\{F > t\} = \{x \in X : F(x) > t\}$ *is open.*

**Proof** Assume that $F$ is not l.s.c. Then there exists a sequence $(x_n)$ converging to $x$ in $X$, and there exists $t \in \mathbb{R}$ such that

$$\lim_{n \to \infty} F(x_n) < t < F(x).$$

Hence for every $r > 0$, $B(x, r) \not\subset \{F > t\}$, and $\{F > t\}$ is not open.

Assume that $\{F > t\}$ is not open. Then there exists a sequence $(x_n)$ converging to $x$ in $X$ such that for every $n$,

$$F(x_n) \leq t < F(x).$$

Hence $\lim_{n \to \infty} F(x_n) < F(x)$ and $F$ is not l.s.c. at $x$.                      $\square$

**Theorem 1.3.13** *Let $X$ be a complete metric space, and let $(F_j : X \to \mathbb{R})_{j \in J}$ be a family of l.s.c. functions such that for every $x \in X$,*

$$\sup_{j \in J} F_j(x) < +\infty. \tag{$*$}$$

*Then there exists a nonempty open subset $V$ of $X$ such that*

$$\sup_{j \in J} \sup_{x \in V} F_j(x) < +\infty.$$

**Proof** By Proposition 1.3.10, the function $F = \sup_{j \in J} F_j$ is l.s.c. The preceding proposition implies that for every $n$, $U_n = \{F > n\}$ is open. By $(*)$, $\bigcap_{n=1}^{\infty} U_n = \phi$. Baire's theorem implies the existence of $n$ such that $U_n$ is not dense. But then $\{F \leq n\}$ contains a nonempty open subset $V$.                      $\square$

**Definition 1.3.14** The characteristic function of $A \subset X$ is defined by

$$\begin{aligned} \chi_A(x) &= 1, \quad x \in A, \\ &= 0, \quad x \in X \setminus A. \end{aligned}$$

**Proposition 1.3.15** *Let $X$ be a metric space and $A \subset X$. Then*

$$A \text{ is open} \iff \chi_A \text{ is l.s.c.;} \quad A \text{ is closed} \iff \chi_A \text{ is u.s.c.}$$

**Definition 1.3.16** Let $S$ be a nonempty subset of a metric space $X$. The distance of $x$ to $S$ is defined on $X$ by $d(x, S) = \inf_{s \in S} d(x, s)$.

**Proposition 1.3.17** *The function "distance to S" is uniformly continuous on X.*

**Proof** Let $x, y \in X$ and $s \in S$. Since $d(x, s) \le d(x, y) + d(y, s)$, we obtain

$$d(x, S) \le \inf_{s \in S} (d(x, y) + d(y, s)) = d(x, y) + d(y, S).$$

We conclude by symmetry that $|d(x, S) - d(y, S)| \le d(x, y)$.                                □

**Definition 1.3.18** Let $Y$ and $Z$ be subsets of a metric space. The distance from $Y$ to $Z$ is defined by $d(Y, Z) = \inf\{d(y, z) : y \in Y, z \in Z\}$.

**Proposition 1.3.19** *Let $Y$ be a compact subset, and let $Z$ be a closed subset of a metric space $X$ such that $Y \cap Z = \phi$. Then $d(Y, Z) > 0$.*

**Proof** Assume that $d(Y, Z) = 0$. Then there exist sequences $(y_n) \subset Y$ and $(z_n) \subset Z$ such that $d(y_n, z_n) \to 0$. By passing, if necessary, to a subsequence, we can assume that $y_n \to y$. But then $d(y, z_n) \to 0$ and $y \in Y \cap Z$.                                □

## 1.4  Convergence

**Definition 1.4.1** Let $X$ be a set and let $Y$ be a metric space. A sequence of mappings $u_n : X \to Y$ converges simply to $u : X \to Y$ if for every $x \in X$,

$$\lim_{n \to \infty} d(u_n(x), u(x)) = 0.$$

The sequence $(u_n)$ converges uniformly to $u$ if

$$\lim_{n \to \infty} \sup_{x \in X} d(u_n(x), u(x)) = 0.$$

*Remarks*

(a) Clearly, uniform convergence implies simple convergence.
(b) The converse is false in general. Let $X = \,]0, 1[$, $Y = \mathbb{R}$, and $u_n(x) = x^n$. The sequence $(u_n)$ converges simply but not uniformly to 0.
(c) We shall prove a partial converse due to Dini.

*Notation* Let $u_n : X \to \overline{\mathbb{R}}$ be a sequence of functions. We write $u_n \uparrow u$ when for every $x$ and for every $n$, $u_n(x) \le u_{n+1}(x)$ and

$$u(x) = \sup_n u_n(x) = \lim_{n \to \infty} u_n(x).$$

We write $u_n \downarrow u$ when for every $x$ and every $n$, $u_{n+1}(x) \le u_n(x)$ and

$$u(x) = \inf_n u_n(x) = \lim_{n \to \infty} u_n(x).$$

**Theorem 1.4.2 (Dini)** *Let $X$ be a compact metric space, and let $u_n : X \to \mathbb{R}$ be a sequence of continuous functions such that*

*(a) $u_n \uparrow u$ or $u_n \downarrow u$;*
*(b) $u : X \to \mathbb{R}$ is continuous.*

*Then $(u_n)$ converges uniformly to $u$.*

**Proof** Assume that

$$0 < \lim_{n \to \infty} \sup_{x \in X} |u_n(x) - u(x)| = \inf_{n \ge 0} \sup_{x \in X} |u_n(x) - u(x)|.$$

There exist $\varepsilon > 0$ and a sequence $(x_n) \subset X$ such that for every $n$,

$$\varepsilon \le |u_n(x_n) - u(x_n)|.$$

By monotonicity, we have for $0 \le m \le n$ that

$$\varepsilon \le |u_m(x_n) - u(x_n)|.$$

By compactness, there exists a sequence $(x_{n_k})$ converging to $x$. By continuity, we obtain for every $m \ge 0$,

$$\varepsilon \le |u_m(x) - u(x)|.$$

But then $(u_n)$ is not simply convergent to $u$. $\qquad\qquad\qquad\qquad\qquad \square$

*Example (Dirichlet Function)* Let us show by an example that two simple limits suffice to destroy *every* point of continuity. Dirichlet's function

$$u(x) = \lim_{m \to \infty} \lim_{n \to \infty} (\cos \pi m! x)^{2n}$$

is equal to 1 when $x$ is rational and to 0 when $x$ is irrational. This function is everywhere discontinuous. Let us prove that uniform convergence preserves continuity.

**Proposition 1.4.3** *Let $X$ and $Y$ be metric spaces, $y \in X$, and $u_n : X \to Y$ a sequence such that*

(a) $(u_n)$ converges uniformly to $u$ on $X$;
(b) for every $n$, $u_n$ is continuous at $y$.

Then $u$ is continuous at $y$.

**Proof** Let $\varepsilon > 0$. By assumption, there exist $n$ and $\delta > 0$ such that

$$\sup_{x \in X} d(u_n(x), u(x)) \le \varepsilon \text{ and } \sup_{x \in B[y,\delta]} d(u_n(x), u_n(y)) \le \varepsilon.$$

Hence for every $x \in B[y, \delta]$,

$$d(u(x), u(y)) \le d(u(x), u_n(x)) + d(u_n(x), u_n(y)) + d(u_n(y), u(y)) \le 3\varepsilon.$$

Since $\varepsilon > 0$ is arbitrary, $u$ is continuous at $y$. $\qquad\qquad\square$

**Definition 1.4.4** Let $X$ be a set and let $Y$ be a metric space. On the space of bounded mappings from $X$ to $Y$,

$$\mathcal{B}(X, Y) = \{u : X \to Y : \sup_{x,y \in X} d(u(x), u(y)) < \infty\},$$

we define the distance of uniform convergence

$$d(u, v) = \sup_{x \in X} d(u(x), v(x)).$$

**Proposition 1.4.5** *Let $X$ be a set and let $Y$ be a complete metric space. Then the space $\mathcal{B}(X, Y)$ is complete.*

**Proof** Assume that $(u_n)$ is such that

$$\lim_{j,k \to \infty} \sup_{x \in X} d(u_j(x), u_k(x)) = 0.$$

Then for every $x \in X$,

$$\lim_{j,k \to \infty} d(u_j(x), u_k(x)) = 0,$$

and the sequence $(u_n(x))$ converges to a limit $u(x)$. Let $\varepsilon > 0$. There exists $m$ such that for $j, k \ge m$ and $x \in X$,

$$d(u_j(x), u_k(x)) \le \varepsilon.$$

By continuity of the distance, we obtain, for $k \ge m$ and $x \in X$,

$$d(u(x), u_k(x)) \le \varepsilon.$$

Hence for $k \geq m$,

$$\sup_{x \in X} d(u(x), u_k(x)) \leq \varepsilon.$$

Since $\varepsilon > 0$ is arbitrary, $(u_n)$ converges uniformly to $u$. It is clear that $u$ is bounded.
□

**Corollary 1.4.6 (Weierstrass Test)** *Let $X$ be a set, and let $u_n : X \to \mathbb{R}$ be a sequence of functions such that*

$$c = \sum_{n=1}^{\infty} \sup_{x \in X} |u_n(x)| < +\infty.$$

*Then the series $\sum_{n=1}^{\infty} u_n$ converges absolutely and uniformly on $X$.*

**Proof** It is clear that for every $x \in X$, $\sum_{n=1}^{\infty} |u_n(x)| \leq c < \infty$. Let us write $v_j = \sum_{n=1}^{j} u_n$. By assumption, we have for $j < k$ that

$$\sup_{x \in X} |v_j(x) - v_k(x)| = \sup_{x \in X} \left| \sum_{n=j+1}^{k} u_n(x) \right| \leq \sum_{n=j+1}^{k} \sup_{x \in X} |u_n(x)| \to 0, \quad j \to \infty.$$

Hence $\lim_{j,k \to \infty} d(v_j, v_k) = 0$, and $(v_j)$ converges uniformly on $X$.
□

*Example (Lebesgue Function)* Let us show by an example that a uniform limit suffices to destroy *every* point of differentiability. Let us define

$$f(x) = \sum_{n=1}^{\infty} \frac{1}{2^n} \sin 2^{n^2} x = \sum_{n=1}^{\infty} u_n(x).$$

Since for every $n$, $\sup_{x \in \mathbb{R}} |u_n(x)| = 2^{-n}$, the convergence is uniform, and the function $f$ is continuous on $\mathbb{R}$. Let $x \in \mathbb{R}$ and $h_{\pm} = \pm \pi/2^{m^2+1}$. A simple computation shows that for $n \geq m + 1$, $u_n(x + h_{\pm}) - u_n(x) = 0$ and

$$\frac{u_m(x + h_{\pm}) - u_m(x)}{h_{\pm}} = \frac{2^{m^2-m+1}}{\pi} [\cos 2^{m^2} x \mp \sin 2^{m^2} x].$$

Let us choose $h = h_+$ or $h = h_-$ such that the absolute value of the expression in brackets is greater than or equal to 1. By the mean value theorem,

$$\left| \sum_{n=1}^{m-1} \frac{u_n(x+h) - u_n(x)}{h} \right| \leq \sum_{n=1}^{m-1} 2^{n^2-n} < 2^{(m-1)^2-(m-1)+1} = 2^{m^2-3m+3}.$$

Hence

$$\frac{2^{m^2-m+1}}{\pi} - 2^{m^2-3m+3} \leq \left| \sum_{n=1}^{m} \frac{u_n(x+h) - u_n(x)}{h} \right| = \left| \frac{f(x+h) - f(x)}{h} \right|,$$

and for every $\varepsilon > 0$,

$$\sup_{0<|h|<\varepsilon} \left| \frac{f(x+h) - f(x)}{h} \right| = +\infty.$$

The Lebesgue function is everywhere continuous and nowhere differentiable. Uniform convergence of the *derivatives* preserves differentiability.

## 1.5 Comments

Our main references on functional analysis are the three classical works

– S. Banach, *Théorie des opérations linéaires* [6],
– F. Riesz and B.S. Nagy, *Leçons d'analyse fonctionnelle* [62],
– H. Brezis, *Analyse fonctionnelle, théorie et applications* [8].

The proof of Ekeland's variational principle [20] in Sect. 1.3 is due to Crandall [21].

The proof of Baire's theorem, Theorem 1.2.13, depends implicitly on the axiom of choice. We need only the following weak form.

**Axiom of Dependent Choices** Let $S$ be a nonempty set, and let $R \subset S \times S$ be such that for each $a \in S$, there exists $b \in S$ satisfying $(a, b) \in S$. Then there is a sequence $(a_n) \subset S$ such that $(a_{n-1}, a_n) \in R$, $n = 1, 2, \ldots$.

We use the notation of Theorem 1.2.13. On

$$S = \{ (m, u, r) : m \in \mathbb{N}, u \in X, r > 0, B(u, r) \subset B \},$$

we define the relation $R$ by

$$\big( (m, u, r), (n, v, s) \big) \in R$$

if and only if $n = m + 1$, $s \le 1/n$, and

$$B[v, s] \subset B(u, r) \cap (\bigcap_{j=1}^{n} U_j).$$

Baire's theorem follows then directly from the axiom of dependent choices.

In 1977, C.E. Blair proved that Baire's theorem implies the axiom of dependent choices, *Bull. Acad. Polon. Sci. Série Sc. Math. Astr. Phys. 25 (1977) 933–934.*

The reader will verify that the axiom of dependent choices is the only principle of choice that we use in this book.

## 1.6   Exercises for Chap. 1

> *La mathématique est une science de problèmes.*
>
> Georges Bouligand

1. Every sequence of real numbers contains a monotonic subsequence. *Hint*: Let

   $$E = \{n \in \mathbb{N} : \text{for every } k \ge n, x_k \le x_n\}.$$

   If $E$ is infinite, $(x_n)$ contains a decreasing subsequence. If $E$ is finite, $(x_n)$ contains an increasing subsequence.
2. Every bounded sequence of real numbers contains a convergent subsequence.
3. Let $(K_n)$ be a decreasing sequence of compact sets and $U$ an open set in a metric space such that $\bigcap_{n=1}^{\infty} K_n \subset U$. Then there exists $n$ such that $K_n \subset U$.
4. Let $(U_n)$ be an increasing sequence of open sets and $K$ a compact set in a metric space such that $K \subset \bigcup_{n=1}^{\infty} U_n$. Then there exists $n$ such that $K \subset U_n$.
5. Define a sequence $(S_n)$ of dense subsets of $\mathbb{R}$ such that $\bigcap_{n=1}^{\infty} S_n = \phi$. Define a family $(U_j)_{j \in J}$ of open dense subsets of $\mathbb{R}$ such that $\bigcap_{j \in J} U_j = \phi$.
6. In a complete metric space, every countable union of closed sets with empty interior has an empty interior. *Hint*: Use Baire's theorem.
7. Dirichlet's function is l.s.c. on $\mathbb{R} \setminus \mathbb{Q}$ and u.s.c. on $\mathbb{Q}$.
8. Let $(u_n)$ be a sequence of functions defined on $[a, b]$ and such that for every $n$,

   $$a \le x \le y \le b \Rightarrow u_n(x) \le u_n(y).$$

Assume that $(u_n)$ converges simply to $u \in C([a, b])$. Then $(u_n)$ converges uniformly to $u$.

9. (Banach fixed-point theorem) Let $X$ be a complete metric space, and let $f : X \to X$ be such that

$$\mathrm{Lip}(f) = \sup\{d(f(x), f(y))/d(x, y) : x, y \in X, x \neq y\} < 1.$$

Then there exists one and only one $x \in X$ such that $f(x) = x$. *Hint*: Consider a sequence defined by $x_0 \in X, x_{n+1} = f(x_n)$.

10. (McShane's extension theorem) Let $Y$ be a subset of a metric space $X$, and let $f : Y \to \mathbb{R}$ be such that

$$\lambda = \mathrm{Lip}(f) = \sup\{|f(x) - f(y)|/d(x, y) : x, y \in Y, x \neq y\} < +\infty.$$

Define on $X$

$$g(x) = \sup\{f(y) - \lambda d(x, y) : y \in Y\}.$$

Then $g\big|_y = f$ and

$$\mathrm{Lip}(g) = \sup\{|g(x) - g(y)/d(x, y) : x, y \in X, x \neq y\} = \mathrm{Lip}(f).$$

11. (Fréchet's extension theorem) Let $Y$ be a dense subset of a metric space $X$, and let $f : Y \to [0, +\infty]$ be an l.s.c. function. Define on $X$

$$g(x) = \inf \left\{ \varliminf_{n \to \infty} f(x_n) : (x_n) \subset Y \text{ and } x_n \to x \right\}.$$

Then $g$ is l.s.c., $g\big|_Y = f$, and for every l.s.c. function $h : X \to [0, +\infty]$ such that $h\big|_Y = f, h \leq g$.

12. Let $X$ be a metric space and $u : X \to [0, +\infty]$ an l.s.c. function such that $u \not\equiv +\infty$. Define

$$u_n(x) = \inf\{u(y) + n\, d(x, y) : y \in X\}.$$

Then $u_n \uparrow u$, and for every $x, y \in X, |u_n(x) - u_n(y)| \leq n\, d(x, y)$.

13. Let $X$ be a metric space and $v : X \to ]-\infty, \infty]$. Then $v$ is l.s.c. if and only if there exists a sequence $(v_n) \subset C(X)$ such that $v_n \uparrow v$. *Hint*: Consider the function $u = \frac{\pi}{2} + \tan^{-1} v$.

14. (Sierpiński, 1921.) Let $X$ be a metric space and $u : X \to \mathbb{R}$. The following properties are equivalent:

(a)  There exists $(u_n) \subset C(X)$ such that for every $x \in X$, $\displaystyle\sum_{n=1}^{\infty} |u_n(x)| < \infty$ and

$$u(x) = \sum_{n=1}^{\infty} u_n(x).$$

(b)  There exists $f, g : X \rightarrow [0, +\infty[$ l.s.c. such that for every $x \in X$, $u(x) = f(x) - g(x)$.

15.  We define

$$X = \{u :]0, 1[ \rightarrow \mathbb{R} : u \text{ is bounded and continuous}\}.$$

We define the distance on $X$ to be

$$d(u, v) = \sup_{x \in ]0, 1[} |u(x) - v(x)|.$$

What are the interior and the closure of

$$Y = \{u \in X : u \text{ is uniformly continuous}\}?$$

# Chapter 2
# The Integral

*Le vrai est simple et clair ; et quand notre manière d'y arriver*
*est embarrassée et obscure, on peut dire qu'elle mène au vrai et*
*n'est pas vraie.*

Fontenelle

## 2.1 The Cauchy Integral

The *Lebesgue integral* is a positive linear functional satisfying the property of monotone convergence. It extends the *Cauchy integral*.

**Definition 2.1.1** Let $\Omega$ be an open subset of $\mathbb{R}^N$. We define

$$C(\Omega) = \{u : \Omega \to \mathbb{R} : u \text{ is continuous}\},$$

$$\mathcal{K}(\Omega) = \{u \in C(\mathbb{R}^N) : \text{spt } u \text{ is a compact subset of } \Omega\}.$$

The *support* of $u$, denoted by spt $u$, is the closure of the set of points at which $u$ is different from 0.

Let $u \in \mathcal{K}(\mathbb{R}^N)$. By definition, there is $R > 1$ such that

$$\text{spt } u \subset \{x \in \mathbb{R}^N : |x|_\infty \leq R - 1\}.$$

Let us define the *Riemann sums* of $u$:

$$S_j = 2^{-jN} \sum_{k \in \mathbb{Z}^N} u(k/2^j).$$

© Springer Nature Switzerland AG 2022
M. Willem, *Functional Analysis*, Cornerstones,
https://doi.org/10.1007/978-3-031-09149-0_2

The factor $2^{-jN}$ is the volume of the cube with side $2^{-j}$ in $\mathbb{R}^N$. Let $C = [0, 1]^N$ and let us define the *Darboux sums* of $u$:

$$A_j = 2^{-jN} \sum_{k \in \mathbb{Z}^N} \min\{u(x) : 2^j x - k \in C\}, \quad B_j = 2^{-jN} \sum_{k \in \mathbb{Z}^N} \max\{u(x) : 2^j x - k \in C\}.$$

Let $\varepsilon > 0$. By uniform continuity, there is $j$ such that $\omega_u(1/2^j) \leq \varepsilon$. Observe that

$$B_j - A_j \leq (2R)^N \varepsilon, \, A_{j-1} \leq A_j \leq S_j \leq B_j \leq B_{j-1}.$$

The *Cauchy integral* of $u$ is defined by

$$\int_{\mathbb{R}^N} u(x)dx = \lim_{j \to \infty} S_j = \lim_{j \to \infty} A_j = \lim_{j \to \infty} B_j.$$

**Theorem 2.1.2**  *The space $\mathcal{K}(\mathbb{R}^N)$ and the Cauchy integral*

$$\Lambda_N : \mathcal{K}(\mathbb{R}^N) \to \mathbb{R} : u \mapsto \int_{\mathbb{R}^N} u \, dx$$

*are such that*

*(a)  for every $u \in \mathcal{K}(\mathbb{R}^N)$, $|u| \in \mathcal{K}(\mathbb{R}^N)$;*
*(b)  for every $u, v \in \mathcal{K}(\mathbb{R}^N)$ and, every $\alpha, \beta \in \mathbb{R}$,*

$$\int_{\mathbb{R}^N} \alpha u + \beta v \, dx = \alpha \int_{\mathbb{R}^N} u \, dx + \beta \int_{\mathbb{R}^N} v \, dx;$$

*(c)  for every $u \in \mathcal{K}(\mathbb{R}^N)$ such that $u \geq 0$, $\int_{\mathbb{R}^N} u \, dx \geq 0$;*

*(d)  for every sequence $(u_n) \subset \mathcal{K}(\mathbb{R}^N)$ such that $u_n \downarrow 0$, $\lim_{n \to \infty} \int_{\mathbb{R}^N} u_n \, dx = 0$.*

**Proof**  Properties (a)–(c) are clear. Property (d) follows from Dini's theorem. By definition, there is $R > 1$ such that

$$\text{spt } u_0 \subset K = \{x \in \mathbb{R}^N : |x|_\infty \leq R - 1\}.$$

By Dini's theorem, $(u_n)$ converges uniformly to 0 on $K$. Hence

$$0 \leq \int_{\mathbb{R}^N} u_n dx \leq (2R)^N \max_{x \in K} u_n(x) \to 0, \quad n \to \infty. \qquad \square$$

It is not always permitted to permute limit and integral.

*Example*  Let us define $(u_n) \subset \mathcal{K}(\mathbb{R})$ by

$$u_n(x) = 2nx(1 - x^2)^{n-1}, \quad 0 < x < 1$$

$$= 0, \qquad\qquad\qquad x \leq 0 \text{ or } x \geq 1,$$

where $n \geq 2$. Then

(a)  for every $x \in \mathbb{R}$, $\lim\limits_{n \to \infty} u_n(x) = 0$;

(b)  for every $n \geq 2$, $\displaystyle\int_{\mathbb{R}} u_n(x)dx = 1$;

(c)  for every $n \geq 2$, spt $u_n = [0, 1]$.

It is easy to verify (a), since, for every $0 < x < 1$,

$$\lim_{n \to \infty} \frac{u_{n+1}(x)}{u_n(x)} = 1 - x^2 < 1,$$

The fundamental theorem of calculus (Theorem 2.2.38) implies (b).

The *(concrete) Lebesgue integral* is the smallest extension of the Cauchy integral satisfying the property of *monotone convergence*

(a)  if $(u_n)$ is a sequence of integrable functions such that $u_n \uparrow u$ and

$$\sup_n \int_{\mathbb{R}^N} u_n \, d\Lambda_N < +\infty,$$

then $u(x) = \lim u_n(x)$ is integrable and

$$\int_{\mathbb{R}^N} u \, d\Lambda_N = \lim_{n \to \infty} \int_{\mathbb{R}^N} u_n \, d\Lambda_N,$$

and the property of *linearity*,

(b)  if $u$ and $v$ are integrable functions and if $\alpha$ and $\beta$ are real numbers, then $\alpha u + \beta v$ is integrable and

$$\int_{\mathbb{R}^N} \alpha u + \beta v \, d\Lambda_N = \alpha \int_{\mathbb{R}^N} u \, d\Lambda_N + \beta \int_{\mathbb{R}^N} v \, d\Lambda_N.$$

By definition, a function $u : \mathbb{R}^N \to ] - \infty, +\infty]$ belongs to $\mathcal{L}^+(\mathbb{R}^N, \Lambda_N)$ if there exists a sequence $(u_n)$ of functions of $\mathcal{K}(\mathbb{R}^N)$ such that $u_n \uparrow u$ and $\sup_n \int_{\mathbb{R}^N} u_n dx < +\infty$. The *integral* of $u$, defined by the formula

$$\int_{\mathbb{R}^N} u \, d\Lambda_N = \lim_{n \to \infty} \int_{\mathbb{R}^N} u_n \, dx,$$

depends only on $u$ and satisfies property (a). It is clear that $\mathcal{K}(\mathbb{R}^N) \subset \mathcal{L}^+(\mathbb{R}^N, \Lambda_N)$. Moreover, for every $u \in \mathcal{K}(\mathbb{R}^N)$,

$$\int_{\mathbb{R}^N} u \, d\Lambda_N = \int_{\mathbb{R}^N} u \, dx.$$

Let $f, g \in \mathcal{L}^+(\mathbb{R}^N, \Lambda_N)$. The difference $f(x) - g(x)$ is well defined except if $f(x) = g(x) = +\infty$. A subset $S$ of $\mathbb{R}^N$ is *negligible* if there exists $h \in \mathcal{L}^+(\mathbb{R}^N, \Lambda_N)$ such that, for every $x \in S$, $h(x) = +\infty$.

By definition, a function $u : \mathbb{R}^N \to [-\infty, +\infty]$ belongs to $\mathcal{L}^1(\mathbb{R}^N, \Lambda_N)$ if there exists $f, g \in \mathcal{L}^+(\mathbb{R}^N, \Lambda_N)$ such that $u = f - g$ except on a negligible subset of $\mathbb{R}^N$. The *integral* of $u$, defined by the formula

$$\int_{\mathbb{R}^N} u \, d\Lambda_N = \int_{\mathbb{R}^N} f \, d\Lambda_N - \int_{\mathbb{R}^N} g \, d\Lambda_N$$

depends only on $u$ and satisfies properties (a) and (b).

After a *descriptive definition* of the (concrete) Lebesgue integral, it was necessary to give a *constructive definition* in order to prove its existence.

The Lebesgue integral will be constructed in an abstract framework, the *elementary integral*, generalizing the Cauchy integral.

## 2.2  The Lebesgue Integral

> *Les inégalités peuvent s'intégrer.*
>
> Paul Lévy

Elementary integrals were defined by Daniell in 1918.

**Definition 2.2.1** An elementary integral on the set $\Omega$ is defined by a vector space $\mathcal{L} = \mathcal{L}(\Omega, \mu)$ of functions from $\Omega$ to $\mathbb{R}$ and by a functional

$$\mu : \mathcal{L} \to \mathbb{R} : u \mapsto \int_\Omega u \, d\mu$$

such that

($\mathcal{J}_1$)  for every $u \in \mathcal{L}$, $|u| \in \mathcal{L}$;
($\mathcal{J}_2$)  for every $u, v \in \mathcal{L}$ and, every $\alpha, \beta \in \mathbb{R}$,

$$\int_\Omega \alpha u + \beta v \, d\mu = \alpha \int_\Omega u \, d\mu + \beta \int_\Omega v \, d\mu;$$

($\mathcal{J}_3$)  for every $u \in \mathcal{L}$ such that $u \geq 0$, $\int_\Omega u \, d\mu \geq 0$.

$(\mathcal{J}_4)$   for every sequence $(u_n) \subset \mathcal{L}$ such that $u_n \downarrow 0$, $\displaystyle \lim_{n \to \infty} \int_\Omega u_n \, d\mu = 0$.

**Proposition 2.2.2** *Let $u, v \in \mathcal{L}$. Then $u^+, u^-, \max(u, v), \min(u, v) \in \mathcal{L}$.*

*Proof* Let us recall that $u^+ = \max(u, 0)$, $u^- = \max(-u, 0)$,

$$\max(u, v) = \frac{1}{2}(u + v) + \frac{1}{2}|u - v|, \quad \min(u, v) = \frac{1}{2}(u + v) - \frac{1}{2}|u - v|. \quad \square$$

**Proposition 2.2.3** *Let $u, v \in \mathcal{L}$ be such that $u \leq v$. Then $\displaystyle \int_\Omega u \, d\mu \leq \int_\Omega v \, d\mu$.*

*Proof* We deduce from $(\mathcal{J}_2)$ and $(\mathcal{J}_3)$ that

$$0 \leq \int_\Omega v - u \, d\mu = \int_\Omega v \, d\mu - \int_\Omega u \, d\mu. \quad \square$$

**Definition 2.2.4** A fundamental sequence is an increasing sequence $(u_n) \subset \mathcal{L}$ such that

$$\lim_{n \to \infty} \int_\Omega u_n d\mu = \sup_n \int_\Omega u_n d\mu < \infty.$$

**Definition 2.2.5** A subset $S$ of $\Omega$ is negligible (with respect to $\mu$) if there is a fundamental sequence $(u_n)$ such that for every $x \in S$, $\displaystyle \lim_{n \to \infty} u_n(x) = +\infty$. A property is true almost everywhere if the set of points of $\Omega$ where it is false is negligible.

Let us justify the definition of a negligible set.

**Proposition 2.2.6** *Let $(u_n)$ be a decreasing sequence of functions of $\mathcal{L}$ such that everywhere $u_n \geq 0$ and almost everywhere, $\displaystyle \lim_{n \to \infty} u_n(x) = 0$. Then*

$$\lim_{n \to \infty} \int_\Omega u_n d\mu = 0.$$

*Proof* Let $\varepsilon > 0$. By assumption, there is a fundamental sequence $(v_n)$ such that if $\displaystyle \lim_{n \to \infty} u_n(x) > 0$, then $\displaystyle \lim_{n \to \infty} v_n(x) = +\infty$. We replace $v_n$ by $v_n^+$, and we multiply by a strictly positive constant such that

$$v_n \geq 0, \qquad \int_\Omega v_n d\mu \leq \varepsilon.$$

We define $w_n = (u_n - v_n)^+$. Then $w_n \downarrow 0$, and we deduce from axiom $(\mathcal{J}_4)$ that

$$0 \leq \lim \int_{\Omega} u_n d\mu \leq \lim \int_{\Omega} w_n + v_n d\mu = \lim \int_{\Omega} w_n d\mu + \lim \int_{\Omega} v_n d\mu$$

$$= \lim \int_{\Omega} v_n d\mu \leq \varepsilon.$$

Since $\varepsilon > 0$ is arbitrary, the proof is complete. □

**Proposition 2.2.7** *Let* $(u_n)$ *and* $(v_n)$ *be fundamental sequences such that almost everywhere,*

$$u(x) = \lim_{n\to\infty} u_n(x) \leq \lim_{n\to\infty} v_n(x) = v(x).$$

*Then*

$$\lim_{n\to\infty} \int_{\Omega} u_n d\mu \leq \lim_{n\to\infty} \int_{\Omega} v_n d\mu.$$

**Proof** We choose $k$ and we define $w_n = (u_k - v_n)^+$. Then $(w_n) \subset \mathcal{L}$ is a decreasing sequence of positive functions such that almost everywhere,

$$\lim w_n(x) = (u_k(x) - v(x))^+ \leq (u(x) - v(x))^+ = 0.$$

We deduce from the preceding proposition that

$$\int_{\Omega} u_k d\mu \leq \lim \int_{\Omega} w_n + v_n \, d\mu = \lim \int_{\Omega} w_n d\mu + \lim \int_{\Omega} v_n d\mu = \lim \int_{\Omega} v_n d\mu.$$

Since $k$ is arbitrary, the proof is complete. □

**Definition 2.2.8** A function $u : \Omega \to \,]-\infty, +\infty]$ belongs to $\mathcal{L}^+ = \mathcal{L}^+(\Omega, \mu)$ if there exists a fundamental sequence $(u_n)$ such that $u_n \uparrow u$. The integral (with respect to $\mu$) of $u$ is defined by

$$\int_{\Omega} u \, d\mu = \lim_{n\to\infty} \int_{\Omega} u_n d\mu.$$

By the preceding proposition, the integral of $u$ is well defined.

**Proposition 2.2.9** *Let* $u, v \in \mathcal{L}^+$ *and* $\alpha, \beta \geq 0$. *Then*

*(a)* $\max(u, v), \min(u, v), u^+ \in \mathcal{L}^+$;

*(b)* $\alpha u + \beta v \in \mathcal{L}^+$ *and* $\int_{\Omega} \alpha u + \beta v \, d\mu = \alpha \int_{\Omega} u \, d\mu + \beta \int_{\Omega} v \, d\mu$;

*(c)* *if* $u \leq v$ *almost everywhere, then* $\int_{\Omega} u \, d\mu \leq \int_{\Omega} v \, d\mu$.

**Proof** Proposition 2.2.7 is equivalent to (c). □

**Lemma 2.2.10 (Monotone Convergence in $\mathcal{L}^+$)** *Let $(u_n) \subset \mathcal{L}^+$ be everywhere (or almost everywhere) increasing and such that*

$$c = \sup_n \int_\Omega u_n d\mu < \infty.$$

*Then $(u_n)$ converges everywhere (or almost everywhere) to $u \in \mathcal{L}^+$ and*

$$\int_\Omega u \, d\mu = \lim_{n\to\infty} \int_\Omega u_n d\mu.$$

*Proof* We consider almost everywhere convergence. For every $k$, there is a fundamental sequence $(u_{k,n})$ such that $u_{k,n} \uparrow u_k$.

The sequence $v_n = \max(u_{1,n}, \dots, u_{n,n})$ is increasing, and almost everywhere,

$$v_n \leq \max(u_1, \dots, u_n) = u_n.$$

Since

$$\int_\Omega v_n d\mu \leq \int_\Omega u_n d\mu \leq c,$$

the sequence $(v_n) \subset \mathcal{L}$ is fundamental. By definition, $v_n \uparrow u$, $u \in \mathcal{L}^+$, and

$$\int_\Omega u \, d\mu = \lim_{n\to\infty} \int_\Omega v_n d\mu.$$

For $k \leq n$, we have almost everywhere that

$$u_{k,n} \leq v_n \leq u_n.$$

Hence we obtain, almost everywhere, that $u_k \leq u \leq \lim_{n\to\infty} u_n$ and

$$\int_\Omega u_k d\mu \leq \int_\Omega u \, d\mu \leq \lim_{n\to\infty} \int_\Omega u_n d\mu.$$

It is easy to conclude the proof. $\qquad\square$

**Theorem 2.2.11** *Every countable union of negligible sets is negligible.*

*Proof* Let $(S_k)$ be a sequence of negligible sets. For every $k$, there exists $v_k \in \mathcal{L}^+$ such that for every $x \in S_k$, $v_k(x) = +\infty$. We replace $v_k$ by $v_k^+$, and we multiply by a strictly positive constant such that

$$v_k \geq 0, \quad \int_\Omega v_k d\mu \leq \frac{1}{2^k}.$$

The sequence $u_n = \sum_{k=1}^{n} v_k$ is increasing and

$$\int_{\Omega} u_n d\mu \le \sum_{k=1}^{n} \frac{1}{2^k} \le 1.$$

Hence $u_n \uparrow u$ and $u \in \mathcal{L}^+$. Since for every $x \in \bigcup_{k=1}^{\infty} S_k$, $u(x) = +\infty$, the set $\bigcup_{k=1}^{\infty} S_k$ is negligible. □

By definition, functions of $\mathcal{L}^+$ are finite almost everywhere. Hence the difference of two functions of $\mathcal{L}^+$ is well defined almost everywhere. Assume that $f, g, v, w \in \mathcal{L}^+$ and that $f - g = v - w$ almost everywhere. Then $f + w = v + g$ almost everywhere and

$$\int_{\Omega} f \, d\mu + \int_{\Omega} w \, dv\mu = \int_{\Omega} f + w \, d\mu = \int_{\Omega} v + g \, d\mu = \int_{\Omega} v \, d\mu + \int_{\Omega} g \, d\mu,$$

so that

$$\int_{\Omega} f \, d\mu - \int_{\Omega} g \, d\mu = \int_{\Omega} v \, d\mu - \int_{\Omega} w \, d\mu.$$

**Definition 2.2.12** A real function $u$ almost everywhere defined on $\Omega$ belongs to $\mathcal{L}^1 = \mathcal{L}^1(\Omega, \mu)$ if there exists $f, g \in \mathcal{L}^+$ such that $u = f - g$ almost everywhere. The integral (with respect to $\mu$) of $u$ is defined by

$$\int_{\Omega} u \, d\mu = \int_{\Omega} f \, d\mu - \int_{\Omega} g \, d\mu.$$

By the preceding computation, the integral is well defined.

**Proposition 2.2.13**

(a) If $u \in \mathcal{L}^1$, then $|u| \in \mathcal{L}^1$.
(b) If $u, v \in \mathcal{L}^1$ and if $\alpha, \beta \in \mathbb{R}$, then $\alpha u + \beta v \in \mathcal{L}^1$ and

$$\int_{\Omega} \alpha u + \beta v \, d\mu = \alpha \int_{\Omega} u \, d\mu + \beta \int_{\Omega} v \, d\mu.$$

(c) If $u \in \mathcal{L}^1$ and if $u \ge 0$ almost everywhere, then $\int_{\Omega} u \, d\mu \ge 0$.

**Proof** Observe that

$$|f - g| = \max(f, g) - \min(f, g).$$ □

**Lemma 2.2.14** *Let* $u \in \mathcal{L}^1$ *and* $\varepsilon > 0$. *Then there exists* $v, w \in \mathcal{L}^+$ *such that* $u = v - w$ *almost everywhere,* $w \geq 0$, *and* $\int_\Omega w \, d\mu \leq \varepsilon$.

***Proof*** By definition, there exists $f, g \in \mathcal{L}^+$ such that $u = f - g$ almost everywhere. Let $(g_n)$ be a fundamental sequence such that $g_n \uparrow g$. Since

$$\int_\Omega g \, d\mu = \lim_{n \to \infty} \int_\Omega g_n d\mu,$$

there exists $n$ such that $\int_\Omega g - g_n \, d\mu \leq \varepsilon$. We choose $w = g - g_n \geq 0$ and $v = f - g_n$. $\qquad\square$

We extend the property of monotone convergence to $\mathcal{L}^1$.

**Theorem 2.2.15 (Levi's Monotone Convergence Theorem)** *Let* $(u_n) \subset \mathcal{L}^1$ *be an almost everywhere increasing sequence such that*

$$c = \sup_n \int_\Omega u_n d\mu < \infty.$$

*Then* $\lim_{n \to \infty} u_n \in \mathcal{L}^1$ *and*

$$\int_\Omega \lim_{n \to \infty} u_n d\mu = \lim_{n \to \infty} \int_\Omega u_n d\mu.$$

***Proof*** After replacing $u_n$ by $u_n - u_0$, we can assume that $u_0 = 0$. By the preceding lemma, for every $k \geq 1$, there exist $v_k, w_k \in \mathcal{L}^+$ such that $w_k \geq 0$, $\int_\Omega w_k d\mu \leq 1/2^k$, and, almost everywhere,

$$u_k - u_{k-1} = v_k - w_k.$$

Since $(u_k)$ is almost everywhere increasing, $v_k \geq 0$ almost everywhere.
We define

$$f_n = \sum_{k=1}^n v_k, \qquad g_n = \sum_{k=1}^n w_k.$$

The sequences $(f_n)$ and $(g_n)$ are almost everywhere increasing, and

$$\int_\Omega g_n d\mu = \sum_{k=1}^n \int_\Omega w_k d\mu \leq \sum_{k=1}^n \frac{1}{2^k} \leq 1, \quad \int_\Omega f_n d\mu = \int_\Omega u_n + g_n d\mu \leq c + 1.$$

Lemma 2.2.10 implies that almost everywhere,

$$\lim_{n \to \infty} f_n = f \in \mathcal{L}^+, \ \lim_{n \to \infty} g_n = g \in \mathcal{L}^+$$

and

$$\int_\Omega f \, d\mu = \lim_{n \to \infty} \int_\Omega f_n d\mu, \ \int_\Omega g \, d\mu = \lim_{n \to \infty} \int_\Omega g \, d\mu.$$

We deduce from Theorem 2.2.11 that almost everywhere,

$$f - g = \lim_{n \to \infty} (f_n - g_n) = \lim_{n \to \infty} u_n.$$

Hence $\lim_{n \to \infty} u_n \in \mathcal{L}^1$ and

$$\int_\Omega \lim_{n \to \infty} u_n d\mu = \int_\Omega f \, d\mu - \int_\Omega g \, d\mu = \lim_{n \to \infty} \int_\Omega f_n - g_n d\mu = \lim_{n \to \infty} \int_\Omega u_n d\mu. \ \square$$

**Theorem 2.2.16 (Fatou's Lemma)** *Let $(u_n) \subset \mathcal{L}^1$ and $f \in \mathcal{L}^1$ be such that*

*(a)* $\sup_n \int_\Omega u_n d\mu < \infty;$
*(b) for every $n$, $f \le u_n$ almost everywhere.*

*Then* $\lim_{n \to \infty} u_n \in \mathcal{L}^1$ *and*

$$\int_\Omega \lim_{n \to \infty} u_n d\mu \le \lim_{n \to \infty} \int_\Omega u_n d\mu.$$

**Proof** We choose $k$, and we define, for $m \ge k$,

$$u_{k,m} = \min(u_k, \dots, u_m).$$

The sequence $(u_{k,m})$ decreases to $v_k = \inf_{n \ge k} u_n$, and

$$\int_\Omega f \, d\mu \le \int_\Omega u_{k,m} d\mu.$$

The preceding theorem, applied to $(-u_{k,m})$, implies that $v_k \in \mathcal{L}^1$ and

$$\int_\Omega v_k d\mu = \lim_{m \to \infty} \int_\Omega u_{k,m} d\mu \le \lim_{m \to \infty} \min_{k \le n \le m} \int_\Omega u_n d\mu = \inf_{n \ge k} \int_\Omega u_n d\mu.$$

The sequence $(v_k)$ increases to $\varliminf_{n \to \infty} u_n$ and

$$\int_\Omega v_k d\mu \le \sup_n \int_\Omega u_n d\mu < \infty.$$

It follows from the preceding theorem that $\varliminf_{n \to \infty} u_n \in \mathcal{L}^1$ and

$$\int_\Omega \varliminf_{n \to \infty} u_n d\mu = \lim_{k \to \infty} \int_\Omega v_k d\mu \le \lim_{k \to \infty} \inf_{n \ge k} \int_\Omega u_n d\mu = \varliminf_{n \to \infty} \int_\Omega u_n d\mu. \quad \square$$

**Theorem 2.2.17 (Lebesgue's Dominated Convergence Theorem)** *Let* $(u_n) \subset \mathcal{L}^1$ *and* $f \in \mathcal{L}^1$ *be such that*

*(a)* $u_n$ *converges almost everywhere;*
*(b)* *for every* $n$, $|u_n| \le f$ *almost everywhere.*

*Then* $\lim_{n \to \infty} u_n \in \mathcal{L}^1$ *and*

$$\int_\Omega \lim_{n \to \infty} u_n d\mu = \lim_{n \to \infty} \int_\Omega u_n d\mu.$$

**Proof** Fatou's lemma implies that $u = \lim_{n \to \infty} u_n \in \mathcal{L}^1$ and

$$2 \int_\Omega f \, d\mu \le \varliminf_{n \to \infty} \int_\Omega 2f - |u_n - u| d\mu = 2 \int_\Omega f \, d\mu - \varlimsup_{n \to \infty} \int_\Omega |u_n - u| d\mu.$$

Hence

$$\lim_{n \to \infty} |\int_\Omega u_n - u \, d\mu| \le \lim_{n \to \infty} \int_\Omega |u_n - u| d\mu = 0. \quad \square$$

**Theorem 2.2.18 (Comparison Theorem)** *Let* $(u_n) \subset \mathcal{L}^1$ *and* $f \in \mathcal{L}^1$ *be such that*

*(a)* $u_n$ *converges almost everywhere to* $u$;
*(b)* $|u| \le f$ *almost everywhere.*

*Then* $u \in \mathcal{L}^1$.

**Proof** We define

$$v_n = \max(\min(u_n, f), -f).$$

The sequence $(v_n) \subset \mathcal{L}^1$ is such that

(a)  $v_n$ converges almost everywhere to $u$;
(b)  for every $n$, $|v_n| \le f$ almost everywhere.

The preceding theorem implies that $u = \lim\limits_{n \to \infty} v_n \in \mathcal{L}^1$.                      $\square$

**Definition 2.2.19**  A real function $u$ defined almost everywhere on $\Omega$ is measurable (with respect to $\mu$) if there exists a sequence $(u_n) \subset \mathcal{L}$ such that $u_n \to u$ almost everywhere. We denote the space of measurable functions (with respect to $\mu$) on $\Omega$ by $\mathcal{M} = \mathcal{M}(\Omega, \mu)$.

**Proposition 2.2.20**

(a)  $\mathcal{L} \subset \mathcal{L}^+ \subset \mathcal{L}^1 \subset \mathcal{M}$.
(b)  If $u \in \mathcal{M}$, then $|u| \in \mathcal{M}$.
(c)  If $u, v \in \mathcal{M}$ and if $\alpha, \beta \in \mathbb{R}$, then $\alpha u + \beta v \in \mathcal{M}$.
(d)  If $u \in \mathcal{M}$ and if, almost everywhere, $|u| \le f \in \mathcal{L}^1$, then $u \in \mathcal{L}^1$.

**Proof**  Property (d) follows from the comparison theorem.                      $\square$

*Notation*  Let $u \in \mathcal{M}$ be such that $u \ge 0$ and $u \notin \mathcal{L}^1$. We write $\displaystyle\int_\Omega u \, d\mu = +\infty$.
Hence the integral of a measurable nonnegative function always exists.

Measurability is preserved by almost everywhere convergence.

**Lemma 2.2.21**  *Let $u \in \mathcal{L}^1$. Then there exists $(u_n) \subset \mathcal{L}$ such that*

(a)  $\displaystyle\int_\Omega |u - u_n| d\mu \to 0, n \to \infty$;
(b)  $u_n \to u, n \to \infty$, a.e. on $\Omega$.

**Proof**  By definition, there exists $f, g \in \mathcal{L}^+$ such that

$$\int_\Omega u \, d\mu = \int_\Omega f \, d\mu - \int_\Omega g \, d\mu, \quad u = f - g, \text{ a.e.}$$

and $(f_n), (g_n) \subset \mathcal{L}$ such that

$$\int_\Omega f \, d\mu = \lim_{n \to \infty} \int_\Omega f_n \, d\mu, \int_\Omega g \, d\mu = \lim_{n \to \infty} \int_\Omega g_n \, d\mu, f_n \uparrow f, g_n \uparrow g, n \to \infty.$$

We define the sequence $(u_n) \subset \mathcal{L}$ by $u_n = f_n - g_n$. Since a.e.

$$|u - u_n| \le f - f_n + g - g_n,$$

it is easy to finish the proof.                      $\square$

**Lemma 2.2.22**  *Let $(u_n) \subset \mathcal{L}^1$ be a sequence converging a.e. to an a.e. finite function $u$. Then $u \in \mathcal{M}$.*

**Proof** The preceding lemma implies the existence of a sequence $(v_n) \subset \mathcal{L}$ such that, for every $n$,

$$\int_\Omega |u_n - v_n| \, d\mu \leq 1/2^n.$$

Since, for every $k$,

$$\int_\Omega \sum_{n=1}^k |u_n - v_n| \, d\mu \leq \sum_{n=1}^k 1/2^n \leq 1,$$

it follows from Levi's monotone convergence theorem that a.e.

$$\sum_{n=1}^\infty |u_n - v_n| < +\infty.$$

Hence we obtain that a.e.

$$u_n - v_n \to 0, n \to \infty,$$

and

$$u = \lim u_n = \lim v_n \in \mathcal{M}. \qquad \square$$

**Lemma 2.2.23**  *Let $(u_n) \subset \mathcal{M}$. Then there exists $f \in \mathcal{L}^+$ such that $f \geq 0$ and a.e.*

$$\sup_n |u_n(x)| > 0 \Rightarrow f(x) > 0. \qquad (*)$$

**Proof** For every $n$, there exists a sequence $(u_{n,j}) \subset \mathcal{L}$ converging a. e. to $u_n$. Let us define $\psi$ on $[0, +\infty[$ by $\psi(0) = 1$ and $\psi(t) = 1/t, t > 0$. By Theorem 2.2.11, the function

$$f = \sum_{n=1}^\infty \sum_{j=1}^\infty 2^{-n-j} \, \psi \left( \int_\Omega |u_{n,j}| \, d\mu \right) |u_{n,j}|$$

satisfies $(*)$. Since, for every $k$,

$$\int_\Omega \sum_{n=1}^k \sum_{j=1}^k 2^{-n-j} \psi \left( \int_\Omega |u_{n,j}| \, d\mu \right) |u_{n,j}| \, d\mu \leq \sum_{n=1}^k \sum_{j=1}^k 2^{-n-j} \leq 1,$$

Lemma 2.2.10 implies that $f \in \mathcal{L}^+$. $\qquad \square$

**Theorem 2.2.24** *Let* $(u_n) \subset M$ *be a sequence converging a. e. to an a.e. finite function u. Then* $u \in M$.

**Proof** Let $f \in \mathcal{L}^+$ be given by the preceding lemma and define,

$$v_n = \max(\min(nf, u_n), -nf).$$

It follows from the comparaison theorem that $(v_n) \subset \mathcal{L}^1$. Since $v_n \to u$ a.e. on $\Omega$, Lemma 2.2.22 implies that $u \in M$.                                                                                                       □

The class of measurable functions is the smallest class containing $\mathcal{L}$ that is closed under almost everywhere convergence.

**Definition 2.2.25** A subset $A$ of $\Omega$ is measurable (with respect to $\mu$) if the characteristic function of $A$ is measurable. The measure of $A$ is defined by

$$\mu(A) = \int_\Omega \chi_A d\mu.$$

**Proposition 2.2.26** *Let A and B be measurable sets, and let* $(A_n)$ *be a sequence of measurable sets. Then* $A \setminus B$, $\displaystyle\bigcup_{n=1}^{\infty} A_n$ *and* $\displaystyle\bigcap_{n=1}^{\infty} A_n$ *are measurable, and*

$$\mu(A \cup B) + \mu(A \cap B) = \mu(A) + \mu(B).$$

*If, moreover, for every n,* $A_n \subset A_{n+1}$, *then*

$$\mu\left(\bigcup_{n=1}^{\infty} A_n\right) = \lim_{n \to \infty} \mu(A_n).$$

*If, moreover,* $\mu(A_1) < \infty$, *and for every n,* $A_{n+1} \subset A_n$, *then*

$$\mu\left(\bigcap_{n=1}^{\infty} A_n\right) = \lim_{n \to \infty} \mu(A_n).$$

**Proof** Observe that

$$\chi_{A \cup B} + \chi_{A \cap B} = \max(\chi_A, \chi_B) + \min(\chi_A, \chi_B) = \chi_A + \chi_B,$$

$$\chi_{A \setminus B} = \chi_A - \min(\chi_A, \chi_B),$$

$$\chi_{\bigcup_{n=1}^{\infty} A_n} = \lim_{n \to \infty} \max(\chi_{A_1}, \dots, \chi_{A_n}),$$

$$\chi_{\bigcap_{n=1}^{\infty} A_n} = \lim_{n \to \infty} \min(\chi_{A_1}, \dots, \chi_{A_n}).$$

The proposition follows then from the preceding theorem and Levi's theorem.      □

**Proposition 2.2.27**  *A subset of $\Omega$ is negligible if and only if it is measurable and its measure is equal to 0.*

**Proof**  Let $A \subset \Omega$ be a negligible set. Since $\chi_A = 0$ almost everywhere, we have by definition that $\chi_A \in \mathcal{L}^1$ and $\mu(A) = \int_\Omega \chi_A d\mu = 0$.

Let $A$ be a measurable set such that $\mu(A) = 0$. For every $n$, $\int_\Omega n\chi_A d\mu = 0$. By Levi's theorem, $u = \lim_{n\to\infty} n\chi_A \in \mathcal{L}^1$. Since $u$ is finite almost everywhere and $u(x) = +\infty$ on $A$, the set $A$ is negligible.                                   $\square$

The hypothesis in the following definition will be used to prove that the set $\{u > t\}$ is measurable when the function $u \geq 0$ is measurable.

**Definition 2.2.28**  A positive measure on $\Omega$ is an elementary integral $\mu : \mathcal{L} \to \mathbb{R}$ on $\Omega$ such that

$(\mathcal{J}_5)$   for every $u \in \mathcal{L}$, $\min(u, 1) \in \mathcal{L}$.

**Proposition 2.2.29**  *Let $\mu$ be a positive measure on $\Omega$, $u \in \mathcal{M}$, and $t \geq 0$. Then $\min(u, t) \in \mathcal{M}$.*

**Proof**  If $t = 0$, $\min(u, 0) = u^+ \in \mathcal{M}$. Let $t > 0$. There is a sequence $(u_n) \subset \mathcal{L}$ converging to $u$ almost everywhere. Then $v_n = t \min(t^{-1}u_n, 1) \in \mathcal{L}$ and $v_n \to \min(u, t)$ almost everywhere.                                          $\square$

**Theorem 2.2.30**  *Let $\mu$ be a positive measure on $\Omega$, and let $u : \Omega \to [0, +\infty]$ be almost everywhere finite. The following properties are equivalent:*

*(a) $u$ is measurable;*
*(b) for every $t \geq 0$, $\{u > t\} = \{x \in \Omega : u(x) > t\}$ is measurable.*

**Proof**  Assume that $u$ is measurable. For every $t \geq 0$ and $n \geq 1$, the preceding proposition implies that

$$u_n = n[\min(u, t + 1/n) - \min(u, t)]$$

is measurable. It follows from Theorem 2.2.24 that

$$\chi_{\{u>t\}} = \lim_{n\to\infty} u_n \in \mathcal{M}.$$

Hence $\{u > t\}$ is measurable.

Assume that $u$ satisfies (b). Let us define, for $n \geq 1$, the function

$$u_n = \frac{1}{2^n} \sum_{k=1}^{\infty} \chi_{\{u>k/2^n\}}.  \tag{$*$}$$

For every $x \in \Omega$, $u(x) - 1/2^n \le u_n(x) \le u(x)$. Hence $(u_n)$ is simply convergent to $u$. Theorem 2.2.24 implies that $(u_n) \subset \mathcal{M}$ and $u \in \mathcal{M}$.                    □

**Corollary 2.2.31** *Let $u, v \in \mathcal{M}$. Then $uv \in \mathcal{M}$.*

**Proof** If $f$ is measurable, then for every $t \ge 0$, the set

$$\{f^2 > t\} = \{|f| > \sqrt{t}\}$$

is measurable. Hence $f^2$ is measurable. We conclude that

$$uv = \frac{1}{4}[(u+v)^2 - (u-v)^2] \in \mathcal{M}.$$                    □

**Definition 2.2.32** A function $u : \Omega \to [0, +\infty]$ is admissible (with respect to the positive measure $\mu$) if $u$ is measurable and if for every $t > 0$,

$$\mu_u(t) = \mu(\{u > t\}) = \mu(\{x \in \Omega : u(x) > t\}) < +\infty.$$

The function $\mu_u$ is the distribution function of $u$.

**Corollary 2.2.33 (Markov Inequality)** *Let $u \in \mathcal{L}^1$, $u \ge 0$. Then $u$ is admissible, and for every $t > 0$,*

$$\mu_u(t) \le t^{-1} \int_{\Omega} u \, d\mu.$$

**Proof** Observe that for every $t > 0$, $v = t\chi_{\{u>t\}} \le u$. By the comparison theorem, $v \in \mathcal{L}^1$ and $\int_{\Omega} v \, d\mu \le \int_{\Omega} u \, d\mu$.                    □

**Corollary 2.2.34 (Cavalieri's Principle)** *Let $u \in \mathcal{L}^1$, $u \ge 0$. Then*

$$\int_{\Omega} u \, d\mu = \int_0^{\infty} \mu_u(t) dt.$$

**Proof** The sequence $(u_n)$ defined by $(*)$ is increasing and converges simply to $u$. The function $\mu_u : ]0, +\infty[ \to [0, +\infty[$ is decreasing. We deduce from Levi's theorem that

$$\int_{\Omega} u \, d\mu = \lim_{n \to \infty} \int_{\Omega} u_n d\mu = \lim_{n \to \infty} \frac{1}{2^n} \sum_{k=1}^{\infty} \mu_u\left(\frac{k}{2^n}\right) = \int_0^{\infty} \mu_u(t) dt.$$                    □

**Definition 2.2.35** Let $\Omega$ be an open subset of $\mathbb{R}^N$. The Lebesgue measure on $\Omega$ is the positive measure defined by the Cauchy integral

$$\Lambda_N : \mathcal{K}(\Omega) \to \mathbb{R} : u \mapsto \int_\Omega u \, dx.$$

We define the functional spaces $\mathcal{L}^+(\Omega) = \mathcal{L}^+(\Omega, \Lambda_N)$ and $\mathcal{L}^1(\Omega) = \mathcal{L}^1(\Omega, \Lambda_N)$. From now on, the Lebesgue integral (with respect to $\Lambda_N$) of $u \in \mathcal{L}^1(\Omega)$ will be denoted by $\int_\Omega u \, dx$. The Lebesgue measure of a $\Lambda_N$-measurable subset $A$ of $\Omega$ is defined by

$$m(A) = \int_\Omega \chi_A dx.$$

Topology is not used in the abstract theory of the *Lebesgue integral*. In contrast, the concrete theory of the *Lebesgue measure* depends on the topology of $\mathbb{R}^N$.

**Theorem 2.2.36** *We consider the Lebesgue measure on $\mathbb{R}^N$.*

*(a) Every open set is measurable, and every closed set is measurable.*
*(b) For every measurable set $A$ of $\mathbb{R}^N$, there exist a sequence $(G_k)$ of open sets of $\mathbb{R}^N$ and a negligible set $S$ of $\mathbb{R}^N$ such that $A \cup S = \bigcap_{k=1}^{\infty} G_k$.*
*(c) For every measurable set $A$ of $\mathbb{R}^N$, there exist a sequence $(F_k)$ of closed sets of $\mathbb{R}^N$ and a negligible set $T$ of $\mathbb{R}^N$ such that $A = \bigcup_{k=1}^{\infty} F_k \cup T$.*

*Proof*

(a) Let $G$ be an open bounded set and define

$$u_n(x) = \min\{1, n \, d(x, \mathbb{R}^N \setminus G)\}. \tag{$*$}$$

Since $(u_n) \subset \mathcal{K}(\mathbb{R}^N)$ and $u_n \to \chi_G$, the set $G$ is measurable. For every open set $G$, $G_n = G \cap B(0, n)$ is measurable. Hence $G = \bigcup_{n=1}^{\infty} G_n$ is measurable. Taking the complement, every closed set is measurable.

(b) Let $A$ be a measurable set of $\mathbb{R}^N$. By definition, there exist a sequence $(u_n) \subset \mathcal{K}(\mathbb{R}^N)$ and a negligible set $R$ of $\mathbb{R}^N$ such that $u_n \to \chi_A$ on $\mathbb{R}^N \setminus R$. There is also $f \in \mathcal{L}^+$ such that $R \subset S = \{f = +\infty\}$. By Proposition 1.3.10, $f$ is l.s.c. Proposition 1.3.12 implies that for every $t \in \mathbb{R}$, $\{f > t\}$ is open. Let us define the open sets

$$U_n = \{u_n > 1/2\} \cup \{f > n\} \quad \text{and} \quad G_k = \bigcup_{n=k}^{\infty} U_n.$$

It is clear that for every $k$, $A \cup S \subset G_k$ and $A \cup S = \bigcap_{k=1}^{\infty} G_k$. Since $S$ is negligible by definition, the proof is complete.

(c) Taking the complement, there exist a sequence $(F_k)$ of closed sets of $\mathbb{R}^N$ and a negligible set $S$ of $\mathbb{R}^N$ such that

$$A \cap (\mathbb{R}^N \setminus S) = \bigcup_{k=1}^{\infty} F_k.$$

It suffices then to define $T = A \cap S$.                                                                          □

**Proposition 2.2.37** *Let $a < b$. Then*

$$m(]a, b[) = m([a, b]) = b - a.$$

*In particular, $m(\{a\}) = 0$, and every countable set is negligible.*

**Proof** Let $(u_n)$ be the sequence defined by (*). It is easy to verify that

$$S_{n,j} = 2^{-j} \sum_{k \in \mathbb{Z}} u_n(k/2^j)$$

satisfies the inequalities

$$b - a - \frac{2}{n} - \frac{1}{2^{j-1}} \leq S_{n,j} \leq b - a + \frac{1}{2^{j-1}}.$$

The definition of the Cauchy integral implies that

$$b - a - \frac{2}{n} \leq \int_{\mathbb{R}} u_n dx = \lim_{j \to \infty} S_{n,j} \leq b - a.$$

Since $u_n \uparrow \chi_{]a,b[}$, it follows from Definition 2.2.8 that

$$m(]a, b[) = \int_{\mathbb{R}} \chi_{]a,b[} dx = \lim_{n \to \infty} \int_{\mathbb{R}} u_n dx = b - a.$$

Since $[a, b] = \bigcap_{n=1}^{\infty} ]a - 1/n, b + 1/n[$, we deduce from Proposition 2.2.26. that

$$m([a, b]) = \lim_{n \to \infty} b - a + 2/n = b - a.$$                                        □

*Notation* If $u$ is integrable on $[a, x[$, we write

$$\int_a^x u(t)dt = \int_{[a,x[} u(t)dt = \int_{\mathbb{R}} \chi_{]a,x[} u(t)dt.$$

The next result, due to A. Cauchy, is the *fundamental theorem of calculus*.

**Theorem 2.2.38**

(a) *Let* $u \in C([a, b])$. *Then for every* $a \leq x \leq b$

$$\frac{d}{dx} \int_a^x u(t)dt = u(x).$$

(b) *Let* $u \in C^1([a, b])$. *Then*

$$\int_a^b u'(t)dt = u(b) - u(a).$$

*Proof*

(a) Let us define $v$ on $[a, b]$ by

$$v(x) = \int_a^x u(t)dt.$$

For every $a \leq x < b$ and for every $0 < \varepsilon \leq b - x$, we have that

$$\left| \frac{v(x + \varepsilon) - v(x)}{\varepsilon} - u(x) \right| = \frac{1}{\varepsilon} \left| \int_x^{x+\varepsilon} (u(t) - u(x))dt \right|$$

$$\leq \frac{1}{\varepsilon} \int_x^{x+\varepsilon} |u(t) - u(x)|dt$$

$$\leq \sup_{x < t < x+\varepsilon} |u(t) - u(x)|.$$

Since $u$ is continuous, we obtain

$$\lim_{\varepsilon \downarrow 0} \left| \frac{v(x + \varepsilon) - v(x)}{\varepsilon} - u(x) \right| = 0.$$

Similarly, for $a < x \leq b$,

$$\lim_{\varepsilon \downarrow 0} \left| \frac{v(x) - v(x - \varepsilon)}{\varepsilon} - u(x) \right| = 0.$$

(b)  Since, for every $a \leq x \leq b$, we have that

$$\frac{d}{dx}\left[\int_a^x u'(t)dt - u(x)\right] = u'(x) - u'(x) = 0,$$

we conclude that

$$\int_a^b u'(t)dt - u(b) = -u(a). \qquad \qquad \qquad \square$$

*Examples*

(a)  Let $\lambda \in \mathbb{R}\backslash\{-1\}$ and $0 < a < b$. Then

$$\int_a^b x^\lambda dx = \left[\frac{x^{\lambda+1}}{\lambda+1}\right]_a^b.$$

(b)  Let $\lambda > -1$ and $b > 0$. For every $n > 1/b$, $X_{[1/n,b[}x^\lambda$ is integrable. Levi's monotone convergence theorem implies that

$$\int_0^b x^\lambda dx = \frac{b^{\lambda+1}}{(\lambda+1)}.$$

(c)  Let $\lambda < -1$ and $a > 0$. For every $n > a$, $X_{[a,n[}x^\lambda$ is integrable. Levi's monotone convergence theorem implies that

$$\int_a^\infty x^\lambda dx = \frac{a^{\lambda+1}}{|\lambda+1|}.$$

(d)  (Cantor sets). Let $0 < \varepsilon \leq 1$ and $(\ell_n) \subset \,]0,1[$ be such that

$$\varepsilon = \sum_{n=0}^\infty 2^n \ell_n.$$

From the interval $C_0 = [0,1]$, remove the open middle interval $J_{0,1}$ of length $\ell_0$. Remove from the two remaining closed intervals the middle open intervals $J_{1,1}$ and $J_{1,2}$ of length $\ell_1$. In general, remove from the $2^n$ remaining closed intervals the middle open intervals $J_{n,1}, \ldots, J_{n,2^n}$ of length $\ell_n$. Define

$$C_{n+1} = C_n \backslash \bigcup_{k=1}^{2^n} J_{n,k}, \quad C = \bigcap_{n=1}^\infty C_n.$$

The set $C$ is the *Cantor set* (corresponding to $(\ell_n)$). Let us describe the fascinating properties of the Cantor set.

*The set $C$ is closed.*   Indeed, each $C_n$ is closed.

*The interior of $C$ is empty.*   Indeed, each $C_n$ consists of $2^n$ closed intervals of equal length, so that $\phi$ is the only open subset in $C$.

*The Lebesgue measure of $C$ is equal to $1 - \varepsilon$.*   By induction, we have for every $n$ that

$$m(C_{n+1}) = 1 - \sum_{j=0}^{n} 2^j \ell_j.$$

Proposition 2.2.26 implies that

$$m(C) = 1 - \sum_{j=0}^{\infty} 2^j \ell_j = 1 - \varepsilon.$$

*The set $C$ is not countable.*   Let $(x_n) \subset C$. Denote by $[a_1, b_1]$ the interval of $C_1$ not containing $x_1$. Denote by $[a_2, b_2]$ the first interval of $C_2 \cap [a_1, b_1]$ not containing $x_2$. In general, let $[a_n, b_n]$ denote the first interval of $C_n \cap [a_{n-1}, b_{n-1}]$ not containing $x_n$. Define $x = \sup_n a_n = \lim_{n \to \infty} a_n$. For every $n$, we have

$$[a_n, b_n] \subset C_n, x_n \notin [a_n, b_n], x \in [a_n, b_n].$$

Hence $x \in C$, and for every $n$, $x_n \neq x$.

For $\varepsilon = 1$, $C$ is not countable and negligible.

Finally, the characteristic function of $C$ is u.s.c., integrable, and discontinuous at every point of $C$.

The first Cantor sets were defined by Smith in 1875, by Volterra in 1881, and by Cantor in 1883.

## 2.3   Multiple Integrals

Fubini's theorem reduces the computation of a double integral to the computation of two simple integrals.

**Definition 2.3.1** Define on $\mathbb{R}$, $f(t) = (1 - |t|)^+$. The family $f_{j,k}(x) = \prod_{n=1}^{N} f(2^j x_n - k_n)$, $j \in \mathbb{N}$, $k \in \mathbb{Z}^N$, is such that $f_{j,k} \in \mathcal{K}(\mathbb{R}^N)$,

$$\operatorname{spt} f_{j,k} = B_\infty[k/2^j, 1/2^j], \quad \sum_{k \in \mathbb{Z}^N} f_{j,k} = 1, \ f_{j,k} \geq 0.$$

**Proposition 2.3.2** *Let $\Omega$ be an open set in $\mathbb{R}^N$ and let $u \in \mathcal{K}(\Omega)$. Then the sequence*

$$u_j = \sum_{k \in \mathbb{Z}^N} u(k/2^j) f_{j,k}$$

*converges uniformly to $u$ on $\Omega$.*

**Proof** Let $\varepsilon > 0$. By uniform continuity, there exists $m$ such that $\omega_u(1/2^m) \leq \varepsilon$. Hence for $j \geq m$,

$$|u(x) - u_j(x)| = |\sum_{k \in \mathbb{Z}^N} (u(x) - u(k/2^j)) f_{j,k}(x)| \leq \varepsilon \sum_{k \in \mathbb{Z}^N} f_{j,k}(x) = \varepsilon. \qquad \square$$

**Proposition 2.3.3** *Let $u \in \mathcal{K}(\mathbb{R}^N)$. Then*

*(a) for every $x_N \in \mathbb{R}$, $u(., x_N) \in \mathcal{K}(\mathbb{R}^{N-1})$;*

*(b) $\int_{\mathbb{R}^{N-1}} u(x', .)dx' \in \mathcal{K}(\mathbb{R})$;*

*(c) $\int_{\mathbb{R}^N} u(x)dx = \int_{\mathbb{R}} dx_N \int_{\mathbb{R}^{N-1}} u(x', x_N)dx'$.*

**Proof** Every restriction of a continuous function is continuous.

Let us define $v(x_N) = \int_{\mathbb{R}^{N-1}} u(x', x_N)dx'$. Lebesgue's dominated convergence theorem implies that $v$ is continuous on $\mathbb{R}$. Since the support of $u$ is a compact subset of $\mathbb{R}^N$, the support of $v$ is a compact subset of $\mathbb{R}$.

We have, for every $j \in \mathbb{N}$ and every $k \in \mathbb{Z}$, by definition of the integral that

$$\int_{\mathbb{R}^N} f_{j,k}(x)dx = \int_{\mathbb{R}} dx_N \int_{\mathbb{R}^{N-1}} f_{j,k}(x', x_N)dx'.$$

Hence for every $j \in \mathbb{N}$,

$$\int_{\mathbb{R}^N} u_j(x)dx = \int_{\mathbb{R}} dx_N \int_{\mathbb{R}^{N-1}} u_j(x', x_N)dx'.$$

There is $R > 1$ such that

$$\text{spt } u \subset \{x \in \mathbb{R}^N : |x|_\infty \le R - 1\}.$$

For every $j \in \mathbb{N}$, by the definition of the integral, we obtain

$$\left| \int_{\mathbb{R}^N} u(x) - u_j(x) dx \right| \le (2R)^N \max_{x \in \mathbb{R}^N} |u(x) - u_j(x)|,$$

and

$$\left| \int_{\mathbb{R}} dx_N \int_{\mathbb{R}^{N-1}} u(x', x_N) - u_j(x', x_N) dx' \right| \le (2R)^N \max_{x \in \mathbb{R}^N} |u(x) - u_j(x)|.$$

It is easy to conclude the proof using the preceding proposition.                □

**Definition 2.3.4**  The elementary integral $\mu$ on $\Omega = \Omega_1 \times \Omega_2$ is the product of the elementary integrals $\mu_1$ on $\Omega_1$ and $\mu_2$ on $\Omega_2$ if for every $u \in \mathcal{L}(\Omega, \mu)$,

(a) $u(., x_2) \in \mathcal{L}(\Omega_1, \mu_1)$ for every $x_2 \in \Omega_2$;

(b) $\int_{\Omega_1} u(x_1, .) d\mu_1 \in \mathcal{L}(\Omega_2, \mu_2)$;

(c) $\int_\Omega u(x_1, x_2) d\mu = \int_{\Omega_2} d\mu_2 \int_{\Omega_1} u(x_1, x_2) d\mu_1.$

We assume that $\mu$ is the product of $\mu_1$ and $\mu_2$.

**Lemma 2.3.5**  Let $u \in \mathcal{L}^+(\Omega, \mu)$. Then

(a) for almost every $x_2 \in \Omega_2$, $u(., x_2) \in \mathcal{L}^+(\Omega_1, \mu_1)$;

(b) $\int_{\Omega_1} u(x_1, .) d\mu_1 \in \mathcal{L}^+(\Omega_2, \mu_2)$;

(c) $\int_\Omega u(x_1, x_2) d\mu = \int_{\Omega_2} d\mu_2 \int_{\Omega_1} u(x_1, x_2) d\mu_1.$

**Proof**  Let $(u_n) \subset \mathcal{L}(\Omega, \mu)$ be a fundamental sequence such that $u_n \uparrow u$. By definition,

$$v_n = \int_\Omega u_n(x_1, .) d\mu_1 \in \mathcal{L}(\Omega_2, \mu_2),$$

and $(v_n)$ is a fundamental sequence. But then $v_n \uparrow v$, $v \in \mathcal{L}^+(\Omega_2, \mu_2)$, and

$$\int_{\Omega_2} v(x_2) d\mu_2 = \lim_{n \to \infty} \int_{\Omega_2} v_n(x_2) d\mu_2.$$

For almost every $x_2 \in \Omega_2$, $v(x_2) \in \mathbb{R}$. In this case, $(u_n(., x_2)) \subset \mathcal{L}(\Omega_1, \mu_1)$ is a fundamental sequence, and $u_n(., x_2) \uparrow u(., x_2)$. Hence $u(., x_2) \in \mathcal{L}^+(\Omega_1, \mu_1)$ and

$$\int_{\Omega_1} u(x_1, x_2)d\mu_1 = \lim_{n\to\infty} \int_{\Omega_1} u_n(x_1, x_2)d\mu_1 = \lim_{n\to\infty} v_n(x_2) = v(x_2).$$

It follows that $\int_{\Omega_1} u(x_1, .)d\mu_1 \in \mathcal{L}^+(\Omega_2, \mu_2)$ and

$$\int_{\Omega} u(x_1, x_2)d\mu = \lim_{n\to\infty} \int_{\Omega} u_n(x_1, x_2)d\mu$$

$$= \lim_{n\to\infty} \int_{\Omega_2} d\mu_2 \int_{\Omega_1} u_n(x_1, x_2)d\mu_1$$

$$= \lim_{n\to\infty} \int_{\Omega_2} v_n(x_2)d\mu_2$$

$$= \int_{\Omega_2} v(x_2)d\mu_2 = \int_{\Omega_2} d\mu_2 \int_{\Omega_1} u(x_1, x_2)d\mu_1. \qquad \square$$

**Lemma 2.3.6** *Let $S \subset \Omega$ be negligible with respect to $\mu$. Then for almost every $x_2 \in \Omega_2$,*

$$S_{x_2} = \{x_1 \in \Omega_1 : (x_1, x_2) \in S\}$$

*is negligible with respect to $\mu_1$.*

**Proof** By assumption, there is $u \in \mathcal{L}^+(\Omega, \mu)$ such that

$$S \subset \{(x_1, x_2) \in \Omega : u(x_1, x_2) = +\infty\}.$$

The preceding lemma implies that for almost every $x_2 \in \Omega_2$,

$$S_{x_2} \subset \{x_1 \in \Omega_1 : u(x_1, x_2) = +\infty\}$$

is negligible with respect to $\mu_1$. $\qquad \square$

**Theorem 2.3.7 (Fubini)** *Let $u \in \mathcal{L}^1(\Omega, \mu)$. Then*

(a) *for almost every $x_2 \in \Omega_2$, $u(., x_2) \in \mathcal{L}^1(\Omega_1, \mu_1)$;*

(b) $\int_{\Omega_1} u(x_1, .)d\mu_1 \in \mathcal{L}^1(\Omega_2, \mu_2)$;

(c) $\int_{\Omega} u(x_1, x_2)d\mu = \int_{\Omega_2} d\mu_2 \int_{\Omega_1} u(x_1, x_2)d\mu_1.$

**Proof** By assumption, there is $f, g \in \mathcal{L}^+(\Omega, \mu)$ such that $u = f - g$ almost everywhere on $\Omega$. By the preceding lemma, for almost every $x_2 \in \Omega_2$,

$$u(x_1, x_2) = f(x_1, x_2) - g(x_1, x_2)$$

almost everywhere on $\Omega_1$. The conclusion follows from Lemma 2.3.5.                    □

The following result provides a way to prove that a function on a product space is integrable.

**Theorem 2.3.8 (Tonelli)**  *Let* $u : \Omega \to [0, +\infty[$ *be such that*

*(a) for every* $n \in \mathbb{N}$, $\min(n, u) \in \mathcal{L}^1(\Omega, \mu)$;

*(b)* $c = \displaystyle\int_{\Omega_2} d\mu_2 \int_{\Omega_1} u(x_1, x_2)d\mu_1 < +\infty.$

*Then* $u \in \mathcal{L}^1(\Omega, \mu)$.

**Proof**  Let us define $u_n = \min(n, u)$. Fubini's theorem implies that

$$\int_{\Omega} u_n(x_1, x_2)d\mu = \int_{\Omega_2} d\mu_2 \int_{\Omega_1} u_n(x_1, x_2)d\mu_1 \le c.$$

The conclusion follows from Levi's monotone convergence theorem.                    □

The following version of Fubini's theorem is due to J.A. Baker.

**Theorem 2.3.9**  *Let* $U$ *be a bounded open subset of* $\mathbb{R}^N$, *and let* $\mu$ *be an elementary integral on the set* $\Omega$. *Assume that* $f \in \mathcal{L}^1(\Omega, \mu)$ *and*

$$F : U \times \Omega \to \mathbb{R} : (x, y) \mapsto F(x, y)$$

*verify*
*(α)* $|F(x, y)| \le f(y)$;
*(β) for* $\mu - $ *almost every* $y \in \Omega$, $F(\cdot, y)$ *is continuous on* $U$;
*(γ) for all* $x \in U$, $F(x, \cdot)$ *is* $\mu -$ *measurable on* $\Omega$.
*   Then*

*(a) the function* $G(x) = \displaystyle\int_{\Omega} F(x, y)d\mu$ *is* $\Lambda_N$-*integrable on* $U$;

*(b) the function* $H(y) = \displaystyle\int_{U} F(x, y)dx$ *is* $\mu$-*integrable on* $\Omega$;

*(c)* $\displaystyle\int_{U} dx \int_{\Omega} F(x, y)d\mu = \int_{\Omega} d\mu \int_{U} F(x, y)dx.$

**Proof**  Let us define on $U \times \Omega$

$$F_j(x, y) = \sum_{k \in \mathbb{Z}^N} F\left(\frac{k}{2^j}, y\right) f_{j,k}(x),$$

where $F(x, y) = 0$ for $x \in \mathbb{R}^N \backslash U$. Assumptions $(\alpha)$ and $(\beta)$ ensure that, for almost every $y \in \Omega$ and for every $x \in U$,

$$\lim_{j\to\infty} F_j(x, y) = F(x, y) \quad , \quad |F_j(x, y)| \le f(y). \quad (*)$$

Lebesgue's dominated convergence theorem, assumption $(\gamma)$, and the continuity of $F_j(\cdot, y)$ imply that

$$G(x) = \lim_{j\to\infty} \int_{\Omega} F_j(x, y)d\mu \quad \text{and} \quad H(y) = \lim_{j\to\infty} \int_{U} F_j(x, y)dx.$$

Let us define

$$G_j(x) = \int_{\Omega} F_j(x, y)d\mu = \sum_{k\in\mathbb{Z}^N} \int_{\Omega} F\left(\frac{k}{2^j}, y\right) d\mu \; f_{j,k}(x)$$

and

$$H_j(y) = \int_{U} F_j(x, y)dx = 2^{-jN} \sum_{k\in\mathbb{Z}^N} F\left(\frac{k}{2^j}, y\right).$$

By definition, for every $j \ge 1$, $G_j$ is continuous and $H_j$ is $\mu$-measurable. It follows from $(*)$ that

$$|G_j(x)| \le \int_{\Omega} f(y)d\mu \quad \text{and} \quad |H_j(y)| \le m(U)f(y).$$

We deduce from Lebesgue's dominated convergence theorem that

$$\int_{U} G(x)dx = \lim_{j\to\infty} \int_{U} G_j(x)dx \quad \text{and} \quad \int_{\Omega} H(y)d\mu = \lim_{j\to\infty} \int_{\Omega} H_j(y)d\mu.$$

Since

$$\int_{U} G_j(x)dx = \int_{U} dx \int_{\Omega} F_j(x, y)d\mu = \int_{\Omega} d\mu \int_{U} F_j(x, y)dx = \int_{\Omega} H_j(y)d\mu,$$

the proof is complete.                                                                            □

## 2.4   Change of Variables

Let $\Omega$ be an open set of $\mathbb{R}^N$, and let $\Lambda_N$ be the Lebesgue measure on $\Omega$. We define

$$\mathcal{L}^+(\Omega) = \mathcal{L}^+(\Omega, \Lambda_N), \mathcal{L}^1(\Omega) = \mathcal{L}^1(\Omega, \Lambda_N).$$

**Definition 2.4.1** Let $\Omega$ and $\omega$ be open. A diffeomorphism is a continuously differentiable bijective mapping $f : \Omega \to \omega$ such that for every $x \in \Omega$,

$$J_f(x) = \det \ f'(x) \neq 0.$$

We assume that $f : \Omega \to \omega$ is a diffeomorphism. The next theorem is proved in Sect. 9.1.

**Theorem 2.4.2** *Let $u \in \mathcal{K}(\omega)$. Then $u(f)|J_f| \in \mathcal{K}(\Omega)$ and*

$$\int_{\Omega} u(f(x))|J_f(x)|dx = \int_{\omega} u(y)dy. \tag{$*$}$$

**Lemma 2.4.3** *Let $u \in \mathcal{L}^+(\omega)$. Then $u(f)|J_f| \in \mathcal{L}^+(\Omega)$, and $(*)$ is valid.*

**Proof** Let $(u_n) \subset \mathcal{K}(\omega)$ be a fundamental sequence such that $u_n \uparrow u$. By the preceding theorem, $v_n = u_n(f)|J_f| \in \mathcal{K}(\Omega)$, and $(v_n)$ is a fundamental sequence. It follows that

$$\int_{\Omega} u(f(x))|J_f(x)|dx = \lim_{n\to\infty} \int_{\Omega} u_n(f(x))|J_f(x)|dx = \lim_{n\to\infty} \int_{\omega} u_n(y)dy = \int_{\omega} u(y)dy.$$

$\square$

**Lemma 2.4.4** *Let $S \subset \omega$ be a negligible set. Then $f^{-1}(S)$ is a negligible set.*

**Proof** By assumption, there is $u \in \mathcal{L}^+(\omega)$ such that

$$S \subset \{y \in \omega : u(y) = +\infty\}.$$

The preceding lemma implies that the set

$$f^{-1}(S) \subset \{x \in \Omega : u(f(x)) = +\infty\}$$

is negligible.

$\square$

**Theorem 2.4.5** *Let $u \in \mathcal{L}^1(\omega)$. Then $u(f)|J_f| \in \mathcal{L}^1(\Omega)$, and $(*)$ is valid.*

**Proof** By assumption, there exists $v, w \in \mathcal{L}^+(\omega)$ such that $u = v - w$ almost everywhere on $\omega$. It follows from the preceding lemma that

$$u(f)|J_f| = v(f)|J_f| - w(f)|J_f|$$

almost everywhere on $\Omega$. It is easy to conclude the proof using Lemma 2.4.3.

$\square$

Let

$$B_N = \{x \in \mathbb{R}^N : |x| < 1\}$$

be the *unit ball* in $\mathbb{R}^N$, and let $V_N = m(B_N)$ be its volume. By the preceding theorem, for every $r > 0$,

$$m(B(0, r)) = \int_{|y|<r} dy = r^N \int_{|x|<1} dx = r^N V_N.$$

We now define *polar coordinates*. Let $N \geq 2$ and $\mathbb{R}_*^N = \mathbb{R}^N \setminus \{0\}$. Let

$$\mathbb{S}^{N-1} = \{\sigma \in \mathbb{R}^N : |\sigma| = 1\}$$

be the *unit sphere* in $\mathbb{R}^N$. The *polar change of variables* is the homeomorphism

$$]0, \infty[ \times \mathbb{S}^{N-1} \longrightarrow \mathbb{R}_*^N : (r, \sigma) \longmapsto r\sigma.$$

**Definition 2.4.6** The surface measure on $\mathbb{S}^{N-1}$ is defined on $C(\mathbb{S}^{N-1})$ by

$$\int_{\mathbb{S}^{N-1}} f(\sigma) d\sigma = N \int_{B_N} f\left(\frac{x}{|x|}\right) dx.$$

Observe that the function $f(x/|x|)$ is bounded and continuous on $B_N \setminus \{0\}$.

Since $\mathbb{S}^{N-1}$ is compact, Dini's theorem implies that the surface measure is a positive measure.

**Lemma 2.4.7** *Let $u \in \mathcal{K}(\mathbb{R}^N)$. Then*

*(a)* *for every $r > 0$, the function $\sigma \mapsto u(r\sigma)$ belongs to $C(\mathbb{S}^{N-1})$;*

*(b)* $\dfrac{d}{dr} \displaystyle\int_{|x|<r} u(x)dx = r^{N-1} \int_{\mathbb{S}^{N-1}} u(r\sigma)d\sigma;$

*(c)* $\displaystyle\int_{\mathbb{R}^N} u(x)dx = \int_0^\infty r^{N-1} dr \int_{\mathbb{S}^{N-1}} u(r\sigma)d\sigma.$

**Proof**

(a)  The restriction of a continuous function is a continuous function.

(b)  Let $w(r) = \displaystyle\int_{|x|<r} u(x)dx$ and $v(r) = \displaystyle\int_{\mathbb{S}^{N-1}} u(r\sigma)d\sigma$, $r > 0$. By definition, we have

$$v(r) = N \int_{B_N} u\left(\frac{r}{|x|}x\right) dx.$$

Choose $r > 0$ and $\varepsilon > 0$. By definition of the modulus of continuity, we have

$$\left| w(r+\varepsilon) - w(r) - \int_{r<|x|<r+\varepsilon} u(rx/|x|)dx \right| = \left| \int_{r<|x|<r+\varepsilon} u(x) - u(rx/|x|)dx \right|$$

$$\leq \omega_u(\varepsilon) V_N [(r+\varepsilon)^N - r^N].$$

The preceding theorem implies that

$$\int_{r<|x|<r+\varepsilon} u(rx/|x|)dx = \int_{|x|<r+\varepsilon} u(rx/|x|)dx - \int_{|x|<r} u(rx/|x|)dx = \frac{(r+\varepsilon)^N - r^N}{N} v(r).$$

Hence we find that

$$\left| w(r+\varepsilon) - w(r) - \frac{(r+\varepsilon)^N - r^N}{N} v(r) \right| \leq \omega_u(\varepsilon) V_N [(r+\varepsilon)^N - r^N],$$

so that

$$\lim_{\substack{\varepsilon \to 0 \\ \varepsilon > 0}} \left| \frac{w(r+\varepsilon) - w(r)}{\varepsilon} - r^{N-1} v(r) \right| = 0.$$

The right derivative of $w$ is equal to $r^{N-1}v$. Similarly, the left derivative of $w$ is equal to $r^{N-1}v$.

(c) The fundamental theorem of calculus implies that for $0 < a < b$,

$$\int_{a<|x|<b} u(x)dx = w(b) - w(a) = \int_a^b v(r)r^{N-1}dr = \int_a^b r^{N-1}dr \int_{\mathbb{S}^{N-1}} u(r\sigma)d\sigma.$$

Taking the limit as $a \to 0$ and $b \to +\infty$, we obtain (c).                    □

**Theorem 2.4.8** *Let $u \in \mathcal{L}^1(\mathbb{R}^N)$. Then*

*(a) for almost every $r > 0$, the function $\sigma \to u(r\sigma)$ belongs to $\mathcal{L}^1(\mathbb{S}^{N-1}, d\sigma)$;*

*(b) the function $r \to \int_{\mathbb{S}^{N-1}} u(r\sigma)d\sigma$ belongs to $\mathcal{L}^1(]0, \infty[, r^{N-1}dr)$;*

*(c) $\int_{\mathbb{R}^N} u(x)dx = \int_0^\infty r^{N-1}dr \int_{\mathbb{S}^{N-1}} u(r\sigma)d\sigma.$*

**Proof** By the preceding theorem, the Lebesgue measure on $\mathbb{R}^N$ is the product of the surface measure on $\mathbb{S}^{N-1}$ and the measure $r^{N-1}dr$ on $]0, \infty[$. It suffices then to use Fubini's theorem.                    □

**Theorem 2.4.9** *The volume $V_N$ is given by the formulas*

$$V_1 = 2, \quad V_2 = \pi \quad and \quad V_N = \frac{2\pi}{N} V_{N-2}.$$

**Proof** Let $N \geq 3$. Fubini's theorem and Theorems 2.4.5 and 2.4.8 imply that

$$
\begin{aligned}
V_N &= \int_{|x|<1} dx \\
&= \int_{x_3^2+\dots+x_N^2<1} dx_3 \dots dx_N \int_{x_1^2+x_2^2<1-(x_3^2+\dots+x_N^2)} dx_1 dx_2 \\
&= \pi \int_{x_3^2+\dots+x_N^2<1} 1 - (x_3^2 + \dots + x_N^2) dx_3 \dots dx_N \\
&= \pi(N-2)V_{N-2} \int_0^1 (1-r^2)r^{N-3} dr = \frac{2\pi}{N} V_{N-2}.
\end{aligned}
$$

$\square$

## 2.5   Comments

The construction of the Lebesgue integral in Chap. 2 follows the article [65] by Roselli and the author. Our source was an outline by Riesz on p. 133 of [62]. However, the space $\mathcal{L}^+$ defined by Riesz is much larger, since it consists of all functions $u$ that are almost everywhere equal to the limit of an almost everywhere increasing sequence $(u_n)$ of elementary functions such that

$$
\sup_n \int_\Omega u_n \, d\mu < \infty.
$$

Using our definition, it is almost obvious that in the case of the concrete Lebesgue integral:

- Every integrable function is almost everywhere equal to the difference of two lower semicontinuous functions.
- The Lebesgue integral is the smallest extension of the Cauchy integral satisfying the properties of monotone convergence and linearity.

Our approach was used in *Analyse Réelle et Complexe* by Golse et al. [30].

Theorem 2.3.9 is due to J.A Baker, *Math. Chronicle 19 (1990) 19–22*.

Lemma 2.4.7 is also due to Baker [4]. The book by Saks [67] is still an excellent reference on integration theory.

The history of integration theory is described in [39, 57]. See also [31] on the life and the work of Émile Borel.

An informal version of the Lebesgue dominated convergence theorem appears (p. 121) in *Théorie du Potentiel Newtonien*, by Henri Poincaré (1899).

## 2.6 Exercises for Chap. 2

1. (Independence of $\mathcal{J}_4$.) The functional defined on

$$\mathcal{L} = \left\{ u : \mathbb{N} \to \mathbb{R} : \lim_{k \to \infty} u(k) \text{ exists} \right\}$$

by $\langle f, u \rangle = \lim_{k \to \infty} u(k)$ satisfies $(\mathcal{J}_{1-2-3})$ but not $\mathcal{J}_4$.

2. (Independence of $\mathcal{J}_5$.) The elementary integral defined on

$$\mathcal{L} = \{ u : [0, 1] \to \mathbb{R} : x \mapsto ax : a \in \mathbb{R} \}$$

by

$$\int u \, d\mu = u(1)$$

is not a positive measure.

3. (Counting measure.) Let $\Omega$ be a set. The elementary integral defined on

$$\mathcal{L} = \{ u : \Omega \to \mathbb{R} : \{ u(x) \neq 0 \} \text{ is finite} \}$$

by

$$\int_{\Omega} u \, d\mu = \sum_{u(x) \neq 0} u(x),$$

satisfies

$$\mathcal{L}^1(\mathbb{N}, \mu) = \left\{ u : \mathbb{N} \to \mathbb{R} : \sum_{n=0}^{\infty} |u(n)| < \infty \right\}$$

and

$$\int_{\mathbb{N}} u \, d\mu = \sum_{n=0}^{\infty} u(n).$$

Prove also that when $\Omega = \mathbb{R}$, the set $\mathbb{R}$ is not measurable.

4. (Axiomatic definition of the Cauchy integral.) Let us recall that $\tau_y u(x) = u(x - y)$. Let $f : \mathcal{K}(\mathbb{R}^N) \to \mathbb{R}$ be a linear functional such that

(a) for every $u \in \mathcal{K}(\mathbb{R}^N)$, $u \geq 0 \Rightarrow \langle f, u \rangle \geq 0$;

(b) for every $y \in \mathbb{R}^N$ and for every $u \in \mathcal{K}(\mathbb{R}^N)$, $\langle f, \tau_y u \rangle = \langle f, u \rangle$.

Then there exists $c \geq 0$ such that for every $u \in \mathcal{K}(\mathbb{R}^N)$, $\langle f, u \rangle = c \int_{\mathbb{R}^N} u \, dx$.

*Hint*: Use Proposition 2.3.2.

5. Let $\mu$ be an elementary integral on $\Omega$. Then the following statements are equivalent:

   (a) $u \in \mathcal{L}^1(\Omega, \mu)$.
   (b) There exists a decreasing sequence $(u_n) \subset \mathcal{L}^+(\Omega, \mu)$ such that almost everywhere, $u = \lim_{n \to \infty} u_n$ and $\inf \int_\Omega u_n d\mu > -\infty$.

6. Let $\Omega = B(0, 1) \subset \mathbb{R}^N$. Then

$$\lambda + N > 0 \iff |x|^\lambda \in \mathcal{L}^1(\Omega), \lambda + N < 0 \iff |x|^\lambda \in \mathcal{L}^1(\mathbb{R}^N \setminus \overline{\Omega}).$$

7. Let $u : \mathbb{R}^2 \to \mathbb{R}$ be such that for every $y \in \mathbb{R}$, $u(., y)$ is continuous and for every $x \in \mathbb{R}$, $u(x, .)$ is continuous. Then $u$ is Lebesgue measurable. *Hint*: Prove the existence of a sequence of continuous functions converging simply to $u$ on $\mathbb{R}^2$.

8. Construct a sequence $(\omega_k)$ of open dense subsets of $\mathbb{R}$ such that $m\left(\bigcap_{k=0}^{\infty} \omega_k\right) = 0$.

   *Hint*: Let $(q_n)$ be an enumeration of $\mathbb{Q}$, and let $I_{n,k}$ be the open interval with center $q_n$ and length $1/2^{n+k}$. Define $\omega_k = \bigcup_{n=0}^{\infty} I_{n,k}$.

9. Prove, using Baire's theorem, that the set of nowhere differentiable functions is dense in $X = C([0, 1])$ with the distance $d(u, v) = \max_{0 \leq x \leq 1} |u(x) - v(x)|$.

   *Hint*: Let $Y$ be the set of functions in $X$ that are differentiable at at least one point, and define, for $n \geq 1$,

$$F_n = \{u \in X : \text{there exists } 0 \leq x \leq 1 \text{ such that,}$$
$$\text{for all } 0 \leq y \leq 1, |u(x) - u(y)| \leq n|x - y|\}.$$

   Since $Y \subset \bigcup_{n=1}^{\infty} F_n$, it suffices to prove that $\bigcap_{n=1}^{\infty} G_n$ is dense in $X$, where $G_n = X \setminus F_n$. By Baire's theorem, it suffices to prove that every $G_n$ is open and dense. It is clear that

$$G_n = \{u \in X : \text{for all } 0 \leq x \leq 1, \text{ there exists } 0 \leq y \leq 1$$
$$\text{such that } n|x - y| < |u(x) - u(y)|\}.$$

   Let $u \in G_n$. The function

$$f(x) = \max\{|u(x) - u(y)| - n(x - y)| : 0 \leq y \leq 1\},$$

is such that

$$\inf_{0 \le x \le 1} f(x) = \min_{0 \le x \le 1} f(x) > 0.$$

It follows that $G_n$ is open.

We use the functions $f_{j,k}$ of Definition 2.3.1. Let $u \in X$ and $\varepsilon > 0$. Define

$$u_j(x) = \sum_{0 \le k \le 2^j} u(k/2^j) f_{j,k}(x),$$

$$g_m(x) = \varepsilon \, d(2^m x, \mathbb{N}).$$

Then for $j$ and $m$ large enough,

$$d(u, u_j) < \varepsilon, \quad u_j + g_m \in G_n.$$

It follows that $G_n$ is dense.

10. Let $\mu$ be a positive measure on the set $\Omega$, and let $u \colon \Omega \to [0, +\infty[$ be a $\mu$-measurable function. Prove that

$$u \in \mathcal{L}^1(\Omega, \mu) \Leftrightarrow \mu_u \in \mathcal{L}^1(]0, +\infty[).$$

In this case

$$\int_\Omega u \, d\mu = \int_0^\infty \mu_u(t) \, dt.$$

11. (Proof of Euler's identity by M. Ivan, 2008).

(a) $\displaystyle \int_{-1}^1 dy \int_{-1}^1 \frac{dx}{1 + 2xy + y^2} = \int_{-1}^1 \frac{\log \frac{1+y}{1-y}}{y} \, dy = 2 \sum_{n=0}^\infty \int_{-1}^1 \frac{y^{2n}}{2n+1} dy$

$$= 4 \sum_{n=0}^\infty \frac{1}{(2n+1)^2}.$$

(b) $\displaystyle \int_{-1}^1 dx \int_{-1}^1 \frac{dy}{1 + 2xy + y^2} = \int_{-1}^1 \frac{\pi}{2\sqrt{1 - x^2}} \, dx = \frac{\pi^2}{2}.$

(c)  The formula $\displaystyle \sum_{n=0}^\infty \frac{1}{(2n+1)^2} = \frac{\pi^2}{8}$ is equivalent to the formula $\displaystyle \sum_{n=1}^\infty \frac{1}{n^2} = \frac{\pi^2}{6}.$

12. Let $u \in C^1(\mathbb{R}^N) \cap \mathcal{K}(\mathbb{R}^N)$. Then

$$u(x) = \frac{1}{N V_N} \int_{\mathbb{R}^N} \frac{\nabla u(x-y) \cdot y}{|y|^N} dy.$$

*Hint:* For every $\sigma \in \mathbb{S}^{N-1}$,

$$u(x) = \int_0^\infty \nabla u(x - r\sigma) \cdot \sigma \, dr.$$

13. The *Newton potential* of the ball $B_R = B(0, R) \subset \mathbb{R}^3$ is defined, for $|y| > R$, by

$$\varphi(y) = \int_{B_R} \frac{dx}{|y - x|}.$$

Since $B_R$ is invariant by rotation, we may assume that $y = (0, 0, a)$, where $a = |y|$. It follows that

$$\varphi(y) = \int_{B_R} \frac{dx}{\sqrt{x_1^2 + x_2^2 + (x_3 - a)^2}}$$

$$= 2\pi \int_{-R}^R dx_3 \int_0^{\sqrt{R^2 - x_3^2}} \frac{r}{\sqrt{r^2 + (x_3 - a)^2}} dr$$

$$= \pi \int_{-R}^R \left( \sqrt{R^2 + a^2 - 2ax_3} - a + x_3 \right) dx_3$$

$$= \frac{4}{3} \pi \frac{R^3}{a} = \frac{4}{3} \pi \frac{R^3}{|y|}.$$

14. The *Newton potential* of the sphere $\mathbb{S}^2$ is defined, for $|y| \neq 1$, by

$$\psi(y) = \int_{\mathbb{S}^2} \frac{d\sigma}{|y - \sigma|}.$$

For $|y| > R$, we have that

$$\frac{4}{3} \pi \frac{R^3}{|y|} = \int_0^R r^2 f(r, y) dr,$$

where

$$f(r, y) = \int_{\mathbb{S}^2} \frac{d\sigma}{|y - r\sigma|}.$$

It follows that

$$4\pi \frac{R^2}{|y|} = R^2 f(R, y).$$

In particular, for $|y| > 1$,

$$\psi(y) = f(1, y) = \frac{4\pi}{|y|}.$$

# Chapter 3
# Norms

## 3.1 Banach Spaces

Since their creation by Banach in 1922, normed spaces have played a central role in functional analysis. Banach spaces are complete normed spaces. Completeness allows one to prove the convergence of a sequence or of a series without using the limit.

**Definition 3.1.1** A norm on a real vector space $X$ is a function

$$X \to \mathbb{R} : u \mapsto ||u||$$

such that

$(\mathcal{N}_1)$   for every $u \in X \setminus \{0\}$, $||u|| > 0$;
$(\mathcal{N}_2)$   for every $u \in X$ and for $\alpha \in \mathbb{R}$, $||\alpha u|| = |\alpha| \, ||u||$;
$(\mathcal{N}_3)$   (Minkowski's inequality) for every $u, v \in X$,

$$||u + v|| \leq ||u|| + ||v||.$$

A (real) normed space is a (real) vector space together with a norm on that space.

*Examples* 1. Let $(X, ||.||)$ be a normed space and let $Y$ be a subspace of $X$. The space $Y$ together with $||.||$ (restricted to $Y$) is a normed space.
2. Let $(X_1, ||.||_1), (X_2, ||.||_2)$ be normed spaces. The space $X_1 \times X_2$ together with

$$||(u_1, u_2)|| = \max(||u_1||_1, ||u_2||_2)$$

is a normed space.

© Springer Nature Switzerland AG 2022
M. Willem, *Functional Analysis*, Cornerstones,
https://doi.org/10.1007/978-3-031-09149-0_3

3. We define the norm on the space $\mathbb{R}^N$ to be

$$|x|_\infty = \max\left\{|x_1|, \ldots, |x_N|\right\}.$$

Every normed space is a metric space.

**Proposition 3.1.2** *Let $X$ be a normed space. The function*

$$X \times X \to \mathbb{R} : (u, v) \mapsto ||u - v||$$

*is a distance on $X$. The following mappings are continuous:*

$$X \to \mathbb{R} : u \mapsto ||u||,$$
$$X \times X \to X : (u, v) \mapsto u + v,$$
$$\mathbb{R} \times X \to X : (\alpha, u) \mapsto \alpha u.$$

*Proof* By $N_1$ and $N_2$,

$$d(u, v) = 0 \Longleftrightarrow u = v, \quad d(u, v) = || - (u - v)|| = ||v - u|| = d(v, u).$$

Finally, by Minkowski's inequality,

$$d(u, w) \le d(u, v) + d(v, w).$$

Since by Minkowski's inequality,

$$\left| ||u|| - ||v|| \right| \le ||u - v||,$$

the norm is continuous on $X$. It is easy to verify the continuity of the sum and of the product by a scalar.                                                                    □

**Definition 3.1.3** Let $X$ be a normed space and $(u_n) \subset X$. The series $\sum\limits_{n=0}^{\infty} u_n$ converges, and its sum is $u \in X$ if the sequence $\sum\limits_{n=0}^{k} u_n$ converges to $u$. We then write $\sum\limits_{n=0}^{\infty} u_n = u$.

The series $\sum\limits_{n=0}^{\infty} u_n$ converges normally if $\sum\limits_{n=0}^{\infty} ||u_n|| < \infty$.

**Definition 3.1.4**  A Banach space is a complete normed space.

**Proposition 3.1.5**  *In a Banach space X, the following statements are equivalent:*

*(a)* $\displaystyle\sum_{n=0}^{\infty} u_n$ *converges;*

*(b)* $\displaystyle\lim_{\substack{j \to \infty \\ j < k}} \sum_{n=j+1}^{k} u_n = 0.$

***Proof*** Define $S_k = \displaystyle\sum_{n=0}^{k} u_n$. Since $X$ is complete, we have

$$(a) \iff \lim_{\substack{j \to \infty \\ j < k}} ||S_k - S_j|| = 0 \iff \lim_{\substack{j \to \infty \\ j < k}} \left\| \sum_{n=j+1}^{k} u_n \right\| = 0 \iff b). \qquad \square$$

**Proposition 3.1.6**  *In a Banach space, every normally convergent series converges.*

***Proof*** Let $\displaystyle\sum_{n=0}^{\infty} u_n$ be a normally convergent series in the Banach space $X$.
Minkowski's inequality implies that for $j < k$,

$$\left\| \sum_{n=j+1}^{k} u_n \right\| \leq \sum_{n=j+1}^{k} ||u_n||.$$

Since the series is normally convergent,

$$\lim_{\substack{j \to \infty \\ j < k}} \sum_{n=j+1}^{k} ||u_n|| = 0.$$

It suffices then to use the preceding proposition. $\qquad \square$

*Examples*  1. The space of bounded continuous functions on the metric space $X$,

$$\mathcal{BC}(X) = \left\{ u \in C(X) : \sup_{x \in X} |u(x)| < \infty \right\},$$

together with the norm

$$||u||_{\infty} = \sup_{x \in X} |u(x)|,$$

is a Banach space. Convergence with respect to $||.||_\infty$ is uniform convergence.

2. Let $\mu$ be a positive measure on $\Omega$. We denote by $L^1(\Omega, \mu)$ the quotient of $\mathcal{L}^1(\Omega, \mu)$ by the equivalence relation "equality almost everywhere". We define the norm

$$||u||_1 = \int_\Omega |u|\, d\mu.$$

Convergence with respect to $||.||_1$ is convergence in mean. We will prove in Sect. 4.2, on Lebesgue spaces, that $L^1(\Omega, \mu)$ is a Banach space.

3. Let $\Lambda_N$ be the Lebesgue measure on the open subset $\Omega$ of $\mathbb{R}^N$. We denote by $L^1(\Omega)$ the space $L^1(\Omega, \Lambda_N)$. Convergence in mean is not implied by simple convergence, and almost everywhere convergence is not implied by convergence in mean.

If $m(\Omega) < \infty$, the comparison theorem implies that for every $u \in \mathcal{BC}(\Omega)$,

$$||u||_1 = \int_\Omega |u|\, dx \le m(\Omega)||u||_\infty.$$

Hence $\mathcal{BC}(\Omega) \subset L^1(\Omega)$, and the canonical injection is continuous, since

$$||u - v||_1 \le m(\Omega)||u - v||_\infty.$$

In order to characterize the convergence in $L^1(\Omega, \mu)$ we shall define the notions of *convergence in measure* and of *equi-integrability*.

We consider a positive measure $\mu$ on $\Omega$. We identify two $\mu$-measurable functions on $\Omega$ when they are $\mu$-almost everywhere equal.

**Definition 3.1.7** A sequence of measurable functions $(u_n)$ converges in measure to a measurable function $u$ if for every $t > 0$,

$$\lim_{n \to \infty} \mu\{|u_n - u| > t\} = 0.$$

**Proposition 3.1.8** *Assume that the sequence $(u_n)$ converges in measure to $u$. Then there exists a subsequence $(u_{n_k})$ converging almost everywhere to $u$ on $\Omega$.*

**Proof** There exists a subsequence $(u_{n_k})$ such that, for every $k$,

$$\mu\{|u_{n_k} - u| > 1/2^k\} \le 1/2^k.$$

Let us define

$$A_k = \{|u_{n_k} - u| > 1/2^k\}, \quad B_k = \Omega \backslash A_k$$

and

$$A = \bigcap_{j=1}^{\infty} \bigcup_{k=j}^{\infty} A_k, \qquad B = \bigcup_{j=1}^{\infty} \bigcap_{k=j}^{\infty} B_k$$

so that $A = \Omega \setminus B$. For every $x \in B$, there exists $j \geq 1$ such that

$$k \geq j \Rightarrow |u_{n_k}(x) - u(x)| \leq 1/2^k.$$

Hence, for every $x \in B$, $\lim_{k \to \infty} u_{n_k}(x) = u(x)$.

Since, for every $j$,

$$\mu(A) \leq \mu \left( \bigcup_{k=j}^{\infty} A_k \right) \leq 2/2^j,$$

we conclude that $\mu(A) = 0$. $\square$

**Proposition 3.1.9** *Let* $(u_n)$ *be a sequence of measurable functions such that*

*(a)* $(u_n)$ *converges to* $u$ *almost everywhere on* $\Omega$,

*(b)* *for every* $\varepsilon > 0$, *there exists a measurable subset* $B$ *of* $\Omega$ *such that* $\mu(B) < \infty$
*and* $\sup_n \int_{\Omega \setminus B} |u_n| d\mu \leq \varepsilon$.

*Then* $(u_n)$ *converges in measure to* $u$.

**Proof** Let $t > 0$ and let $\varepsilon > 0$. By assumption (b) there exists a measurable subset $B$ of $\Omega$ such that $\mu(B) < \infty$ and $\sup_n \int_{\Omega \setminus B} |u_n| d\mu \leq \varepsilon t/3$. It follows from Fatou's lemma that $\int_{\Omega \setminus B} |u| d\mu \leq \varepsilon t/3$. Lebesgue's dominated convergence theorem implies the existence of $m$ such that

$$n \geq m \Rightarrow \int_B \chi_{|u_n - u| > t} \, d\mu \leq \varepsilon/3.$$

We conclude using Markov's inequality that, for $n \geq m$,

$$\mu \{|u_n - u| > t\} \leq \int_B \chi_{|u_n - u| > t} \, d\mu + \frac{1}{t} \int_{\Omega \setminus B} |u_n - u| d\mu$$

$$\leq \frac{\varepsilon}{3} + \frac{1}{t} \int_{\Omega \setminus B} |u_n| d\mu + \frac{1}{t} \int_{\Omega \setminus B} |u| d\mu \leq \varepsilon. \qquad \square$$

**Proposition 3.1.10**  *Let $u \in L^1(\Omega, \mu)$ and let $\varepsilon > 0$. Then*

*(a)  there exists $\delta > 0$ such that, for every measurable subset $A$ of $\Omega$*

$$\mu(A) \leq \delta \Rightarrow \int_A |u| d\mu \leq \varepsilon \,;$$

*(b)  there exists a measurable subset $B$ of $\Omega$ such that $\mu(B) < \infty$ and $\int_{\Omega \setminus B} |u| d\mu \leq \varepsilon$.*

***Proof***     (a) By Lebesgue's dominated convergence theorem, there exists $m$ such that

$$\int_{|u|>m} |u| d\mu \leq \varepsilon/2.$$

Let $\delta = \varepsilon/(2m)$. For every measurable subset $A$ of $\Omega$ such that $\mu(A) \leq \delta$, we have that

$$\int_A |u| d\mu \leq m\mu(A) + \int_{|u|>m} |u| d\mu \leq \varepsilon.$$

(b)  By Lebesgue's dominated convergence theorem, there exists $n$ such that

$$\int_{|u| \leq 1/n} |u| d\mu \leq \varepsilon.$$

The set $B = \{|u| > 1/n\}$ is such that $\mu(B) < \infty$ and $\int_{\Omega \setminus B} |u| d\mu \leq \varepsilon$.     □

**Definition 3.1.11**  A subset $S$ of $L^1(\Omega, \mu)$ is equi-integrable if

(a)  for every $\varepsilon > 0$, there exists $\delta > 0$ such that, for every measurable subset $A$ of $\Omega$ satisfying $\mu(A) \leq \delta$, $\sup\limits_{u \in S} \int_A |u| d\mu \leq \varepsilon$,

(b)  for every $\varepsilon > 0$, there exists a measurable subset $B$ of $\Omega$ such that $\mu(B) < \infty$ and $\sup\limits_{u \in S} \int_{\Omega \setminus B} |u| d\mu \leq \varepsilon$.

**Theorem 3.1.12 (Vitali)**  *Let $(u_n) \subset L^1(\Omega, \mu)$ and let $u$ be a measurable function. Then the following properties are equivalent:*

*(a)  $\|u_n - u\|_1 \to 0$, $n \to \infty$,*

*(b)  $(u_n)$ converges in measure to $u$ and $\{u_n : n \in \mathbb{N}\}$ is equi-integrable.*

***Proof***  Assume that (a) is satisfied. Markov's inequality implies that, for every $t > 0$,

$$\mu\{|u_n - u| > t\} \leq \frac{1}{t}\|u_n - u\|_1 \to 0, n \to \infty.$$

Let $\varepsilon > 0$. There exists $m$ such that

$$n \geq m \Rightarrow \|u_n - u\|_1 \leq \varepsilon/2.$$

In particular, for every measurable subset $A$ of $\Omega$ and for every $n \geq m$,

$$\int_A |u_n|d\mu \leq \int_A |u|d\mu + \int_A |u_n - u|d\mu \leq \int_A |u|d\mu + \varepsilon/2.$$

Proposition 3.1.10 implies the existence of $\delta > 0$ such that, for every measurable subset $A$ of $\Omega$,

$$\mu(A) \leq \delta \Rightarrow \int_A \sup\Big(2|u|, |u_1|, ..., |u_{m-1}|\Big)d\mu \leq \varepsilon.$$

We conclude that, for every measurable subset $A$ of $\Omega$,

$$\mu(A) \leq \delta \Rightarrow \sup_n \int_A |u_n|d\mu \leq \varepsilon.$$

Similarly, Proposition 3.1.10 implies the existence of a measurable subset $B$ of $\Omega$ such that $\mu(B) < \infty$ and

$$\int_{\Omega\setminus B} \sup\Big(2|u|, |u_1|, ..., |u_{m-1}|\Big)d\mu \leq \varepsilon.$$

We conclude that $\sup_n \int_{\Omega\setminus B} |u_n|d\mu \leq \varepsilon.$

Assume now that (b) is satisfied. Let $\varepsilon > 0$. By assumption, there exists $\delta > 0$ such that, for every measurable subset $A$ of $\Omega$,

$$\mu(A) \leq \delta \Rightarrow \sup_n \int_A |u_n|d\mu \leq \varepsilon,$$

and there exists a measurable subset $B$ of $\Omega$ such that $\mu(B) < \infty$ and

$$\sup_n \int_{\Omega\setminus B} |u_n|d\mu \leq \varepsilon.$$

We assume that $\mu(B) > 0$. The case $\mu(B) = 0$ is simpler. Since $(u_n)$ converges in measure to $u$, Proposition 3.1.8 implies the existence of a subsequence $(u_{n_k})$ such that $u_{n_k} \to u$ almost everywhere on $\Omega$. It follows from Fatou's lemma that, for every measurable subset $A$ of $\Omega$,

$$\mu(A) \leq \delta \Rightarrow \int_A |u| d\mu \leq \varepsilon,$$

and that

$$\int_{\Omega \setminus B} |u| d\mu \leq \varepsilon.$$

There exists also $m$ such that

$$n \geq m \Rightarrow \mu\{|u_n - u| > \varepsilon/\mu(B)\} \leq \delta.$$

Let us define $A_n = \{|u_n - u| > \varepsilon/\mu(B)\}$, so that, for $n \geq m$, $\mu(A_n) \leq \delta$. For every $n \geq m$, we obtain

$$\int_{\Omega} |u_n - u| d\mu \leq \int_{\Omega \setminus B} |u_n| + |u| d\mu + \int_{A_n} |u_n| + |u| d\mu + \int_{B \setminus A_n} |u_n - u| d\mu$$

$$\leq 4\varepsilon + \int_{B \setminus A_n} \varepsilon/\mu(B) d\mu \leq 5\varepsilon.$$

Since $\varepsilon > 0$ is arbitrary, the proof is complete. □

The following characterization is due to de la Vallée Poussin.

**Theorem 3.1.13** *Let* $S \subset L^1(\Omega, \mu)$ *be such that* $c = \sup_{u \in S}\|u\|_1 < +\infty$. *The following properties are equivalent:*

*(a) for every* $\varepsilon > 0$ *there exists* $\delta > 0$ *such that, for every measurable subset A of* $\Omega$

$$\mu(A) \leq \delta \Rightarrow \sup_{u \in S} \int_A |u| d\mu \leq \varepsilon,$$

*(b) there exists a strictly increasing convex function* $F : [0, +\infty[ \rightarrow [0, +\infty[$ *such that*

$$\lim_{t \to \infty} F(t)/t = +\infty, \quad M = \sup_{u \in S} \int_{\Omega} F(|u|) d\mu < +\infty.$$

**Proof** Since, by Markov's inequality

$$\sup_{u \in S} \mu\{|u| > t\} \leq c/t,$$

assumption (a) implies the existence of a sequence $(n_k)$ of integers such that, for every $k$,

$$n_k < n_{k+1} \quad \text{and} \quad \sup_{u \in S} \int_{|u| > n_k} |u| d\mu \le 1/2^k.$$

Let us define $F(t) = t + \sum_{k=1}^{\infty} (t - n_k)^+$. It is clear that $F$ is strictly increasing and convex. Moreover, for every $j$,

$$t > 2n_{2j} \Rightarrow j \le F(t)/t$$

and, for every $u \in S$, by Levi's theorem,

$$\int_{\Omega} F(|u|) d\mu = \int_{\Omega} |u| d\mu + \sum_{k=1}^{\infty} \int_{\Omega} (|u| - n_k)^+ d\mu \le \int_{\Omega} |u| d\mu + \sum_{k=1}^{\infty} \int_{|u| > n_k} |u| d\mu \le c + 1,$$

so that $S$ satisfies (b).

Assume now that $S$ satisfies (b). Let $\varepsilon > 0$. There exists $s > 0$ such that for every $t \ge s$, $F(t)/t \ge 2M/\varepsilon$. Hence for every $u \in S$ we have that

$$\int_{|u| > s} |u| d\mu \le \frac{\varepsilon}{2M} \int_{|u| > s} F(|u|) d\mu \le \varepsilon/2.$$

We choose $\delta = \varepsilon/(2s)$. For every measurable subset $A$ of $\Omega$ such that $\mu(A) \le \delta$ and for every $u \in S$, we obtain

$$\int_A |u| d\mu \le s\mu(A) + \int_{|u| > s} |u| d\mu \le \varepsilon. \qquad \square$$

## 3.2 Continuous Linear Mappings

> On a le droit de faire la théorie générale des opérations sans définir l'opération que l'on considère, de même qu'on fait la théorie de l'addition sans définir la nature des termes à additionner.
>
> Henri Poincaré

In general, linear mappings between normed spaces are not continuous.

**Proposition 3.2.1** *Let X and Y be normed spaces and $A : X \to Y$ a linear mapping. The following properties are equivalent:*

*(a)  A is continuous;*

*(b)  $c = \displaystyle\sup_{\substack{u \in X \\ u \neq 0}} \dfrac{||Au||}{||u||} < \infty.$*

**Proof** If $c < \infty$, we obtain

$$||Au - Av|| = ||A(u - v)|| \leq c||u - v||.$$

Hence $A$ is continuous.

If $A$ is continuous, there exists $\delta > 0$ such that for every $u \in X$,

$$||u|| = ||u - 0|| \leq \delta \Rightarrow ||Au|| = ||Au - A0|| \leq 1.$$

Hence for every $u \in X \setminus \{0\}$,

$$||Au|| = \frac{||u||}{\delta} ||A\left(\frac{\delta}{||u||}u\right)|| \leq \frac{||u||}{\delta}. \qquad \square$$

**Proposition 3.2.2** *The function*

$$||A|| = \sup_{\substack{u \in X \\ u \neq 0}} \frac{||Au||}{||u||} = \sup_{\substack{u \in X \\ ||u||=1}} ||Au||$$

*defines a norm on the space $\mathcal{L}(X, Y) = \{A : X \to Y : A \text{ is linear and continuous}\}$.*

**Proof** By the preceding proposition, if $A \in \mathcal{L}(X, Y)$, then $0 \leq ||A|| < \infty$. If $A \neq 0$, it is clear that $||A|| > 0$. It follows from axiom $\mathcal{N}_2$ that

$$||\alpha A|| = \sup_{\substack{u \in X \\ ||u|| = 1}} ||\alpha Au|| = \sup_{\substack{u \in X \\ ||u|| = 1}} |\alpha| \, ||Au|| = |\alpha| \, ||A||.$$

It follows from Minkowski's inequality that

$$||A + B|| = \sup_{\substack{u \in X \\ ||u|| = 1}} ||Au + Bu|| \leq \sup_{\substack{u \in X \\ ||u|| = 1}} (||Au|| + ||Bu||) \leq ||A|| + ||B||. \qquad \square$$

**Proposition 3.2.3 (Extension by density)** *Let Z be a dense subspace of a normed space X, Y a Banach space, and $A \in \mathcal{L}(Z, Y)$. Then there exists a unique mapping $B \in \mathcal{L}(X, Y)$ such that $B\big|_Z = A$. Moreover, $||B|| = ||A||$.*

**Proof** Let $u \in X$. There exists a sequence $(u_n) \subset Z$ such that $u_n \to u$. The sequence $(Au_n)$ is a Cauchy sequence, since

$$||Au_j - Au_k|| \leq ||A|| \, ||u_j - u_k|| \to 0, \quad j, k \to \infty$$

by Proposition 1.2.3. We denote by $f$ its limit. Let $(v_n) \subset Z$ be such that $v_n \to u$. We have

$$||Av_n - Au_n|| \leq ||A|| \, ||v_n - u_n|| \leq ||A|| \, (||v_n - u|| + ||u - u_n||) \to 0, \quad n \to \infty.$$

Hence $Av_n \to f$, and we define $Bu = f$. By Proposition 3.1.2, $B$ is linear. Since for every $n$,

$$||Au_n|| \leq ||A|| \, ||u_n||,$$

we obtain by Proposition 3.1.2 that

$$||Bu|| \leq ||A|| \, ||u||.$$

Hence $B$ is continuous and $||B|| \leq ||A||$. It is clear that $||A|| \leq ||B||$. Hence $||A|| = ||B||$.

If $C \in \mathcal{L}(X, Y)$ is such that $C|_Z = A$, we obtain

$$Cu = \lim_{n \to \infty} Cu_n = \lim_{n \to \infty} Au_n = \lim_{n \to \infty} Bu_n = Bu. \qquad \square$$

**Proposition 3.2.4** *Let $X$ and $Y$ be normed spaces, and let $(A_n) \subset \mathcal{L}(X, Y)$ and $A \in \mathcal{L}(X, Y)$ be such that $||A_n - A|| \to 0$. Then $(A_n)$ converges simply to $A$.*

**Proof** For every $u \in X$, we have

$$||A_n u - Au|| = ||(A_n - A)u|| \leq ||A_n - A|| \, ||u||. \qquad \square$$

**Proposition 3.2.5** *Let $Z$ be a dense subset of a normed space $X$, let $Y$ be a Banach space, and let $(A_n) \subset \mathcal{L}(X, Y)$ be such that*

*(a) $c = \sup_n ||A_n|| < \infty$;*
*(b) for every $v \in Z$, $(A_n v)$ converges.*

*Then $A_n$ converges simply to $A \in \mathcal{L}(X, Y)$, and*

$$||A|| \leq \varliminf_{n \to \infty} ||A_n||.$$

**Proof** Let $u \in X$ and $\varepsilon > 0$. By density, there exists $v \in B(u, \varepsilon) \cap Z$. Since $(A_n v)$ converges, Proposition 1.2.3 implies the existence of $n$ such that

$$j, k \geq n \Rightarrow \|A_j v - A_k v\| \leq \varepsilon.$$

Hence for $j, k \geq n$, we have

$$\begin{aligned}
\|A_j u - A_k u\| &\leq \|A_j u - A_j v\| + \|A_j v - A_k v\| + \|A_k v - A_k u\| \\
&\leq 2c \|u - v\| + \varepsilon \\
&= (2c + 1)\varepsilon.
\end{aligned}$$

The sequence $(A_n u)$ is a Cauchy sequence, since $\varepsilon > 0$ is arbitrary. Hence $(A_n u)$ converges to a limit $Au$ in the complete space $Y$. It follows from Proposition 3.1.2 that $A$ is linear and that

$$\|Au\| = \lim_{n \to \infty} \|A_n u\| \leq \lim_{n \to \infty} \|A_n\| \, \|u\|.$$

But then $A$ is continuous and $\|A\| \leq \lim_{n \to \infty} \|A_n\|$. $\qquad\qquad\qquad\qquad\qquad\square$

**Theorem 3.2.6 (Banach–Steinhaus theorem)** *Let $X$ be a Banach space, let $Y$ be a normed space, and let $(A_n) \subset \mathcal{L}(X, Y)$ be such that for every $u \in X$,*

$$\sup_n \|A_n u\| < \infty.$$

*Then*

$$\sup_n \|A_n\| < \infty.$$

**First Proof** Theorem 1.3.13 applied to the sequence $F_n : u \mapsto \|A_n u\|$ implies the existence of a ball $B(v, r)$ such that

$$c = \sup_n \sup_{u \in B(v,r)} \|A_n u\| < \infty.$$

It is clear that for every $y, z \in Y$,

$$\|y\| \leq \max\{\|z + y\|, \|z - y\|\}. \tag{$*$}$$

Hence for every $n$ and for every $w \in B(0, r)$, $\|A_n w\| \leq c$, so that

$$\sup_n \|A_n\| \leq c/r.$$

**Second Proof** Assume to obtain a contradiction that $\sup_n \|A_n\| = +\infty$. By considering a subsequence, we assume that $n \, 3^n \leq \|A_n\|$. Let us define inductively a sequence $(u_n)$. We choose $u_0 = 0$. There exists $v_n$ such that $\|v_n\| = 3^{-n}$ and $\frac{3}{4} 3^{-n} \|A_n\| \leq \|A_n v_n\|$. By $(*)$, replacing if necessary $v_n$ by $-v_n$, we obtain

$$\frac{3}{4} 3^{-n} \|A_n\| \leq \|A_n v_n\| \leq \|A_n (u_{n-1} + v_n)\|.$$

We define $u_n = u_{n-1} + v_n$, so that $\|u_n - u_{n-1}\| = 3^{-n}$. It follows that for every $k \geq n$,

$$\|u_k - u_n\| \leq 3^{-n}/2.$$

Hence $(u_n)$ is a Cauchy sequence that converges to $u$ in the complete space $X$. Moreover,

$$\|u - u_n\| \leq 3^{-n}/2.$$

We conclude that

$$\|A_n u\| \geq \|A_n u_n\| - \|A_n (u_n - u)\|$$

$$\geq \|A_n\| \left[ \frac{3}{4} 3^{-n} - \|u_n - u\| \right]$$

$$\geq n \, 3^n \left[ \frac{3}{4} 3^{-n} - \frac{1}{2} 3^{-n} \right] = n/4. \qquad \square$$

**Corollary 3.2.7** *Let $X$ be a Banach space, $Y$ a normed space, and $(A_n) \subset \mathcal{L}(X, Y)$ a sequence converging simply to $A$. Then $(A_n)$ is bounded, $A \in \mathcal{L}(X, Y)$, and*

$$\|A\| \leq \varliminf_{n \to \infty} \|A_n\|.$$

**Proof** For every $u \in X$, the sequence $(A_n u)$ is convergent, hence bounded, by Proposition 1.2.3. The Banach–Steinhaus theorem implies that $\sup_n \|A_n\| < \infty$. It follows from Proposition 3.1.2 that $A$ is linear and

$$\|Au\| = \lim_{n \to \infty} \|A_n u\| \leq \varliminf_{n \to \infty} \|A_n\| \|u\|,$$

so that $A$ is continuous and $\|A\| \leq \varliminf_{n \to \infty} \|A_n\|$. $\qquad \square$

The preceding corollary explains why every natural linear mapping defined on a Banach space is continuous.

*Examples (Convergence of functionals)* We define the linear continuous functionals $f_n$ on $L^1(]0, 1[)$ to be

$$\langle f_n, u \rangle = \int_0^1 u(x)x^n\,dx.$$

Since for every $u \in L^1(]0, 1[)$ such that $||u||_1 = 1$, we have

$$|\langle f_n, u \rangle| < \int_0^1 |u(x)|dx = 1,$$

it is clear that

$$||f_n|| = \sup_{\substack{u \in L^1 \\ ||u||_1 = 1}} |\langle f_n, u \rangle| \le 1.$$

Choosing $v_k(x) = (k+1)x^k$, we obtain

$$\lim_{k \to \infty} \langle f_n, v_k \rangle = \lim_{k \to \infty} \frac{k+1}{k+n+1} = 1.$$

It follows that $||f_n|| = 1$, and for every $u \in L^1(]0, 1[)$ such that $||u||_1 = 1$,

$$|\langle f_n, u \rangle| < ||f_n||.$$

Lebesgue's dominated convergence theorem implies that $(f_n)$ converges simply to $f = 0$. Observe that

$$||f|| < \lim_{n \to \infty} ||f_n||.$$

**Definition 3.2.8** A seminorm on a real vector space $X$ is a function $F: X \to [0, +\infty[$ such that

(a) for every $u \in X$ and for every $\alpha \in \mathbb{R}$, $F(\alpha u) = |\alpha|F(u)$, (positive homogeneity);
(b) for every $u, v \in X$, $F(u + v) \le F(u) + F(v)$, (subadditivity).

*Examples* (a) Any norm is a seminorm.
(b) Let $X$ be a real vector space, $Y$ a normed space, and $A: X \to Y$ a linear mapping. The function $F$ defined on $X$ by $F(u) = ||Au||$ is a seminorm.
(c) Let $X$ be a normed space, $Y$ a real vector space, and $A: X \to Y$ a surjective linear mapping. The function $F$ defined on $Y$ by

$$F(v) = \inf\left\{\|u\| : Au = v\right\}$$

is a seminorm.

**Proposition 3.2.9** *Let $F$ be a seminorm defined on a normed space $X$. The following properties are equivalent*

*(a)  $F$ is continuous;*
*(b)  $c = \sup\limits_{\substack{u \in X \\ \|u\|=1}} F(u) < \infty.$*

**Proof** If $F$ satisfies (b), then

$$\left|F(u) - F(v)\right| \le F(u - v) \le c\|u - v\|,$$

so that $F$ is continuous.

It is easy to prove that the continuity of $F$ at 0 implies (b).                            □

Let $F$ be a seminorm on the normed space $X$ and consider a convergent series $\sum\limits_{k=1}^{\infty} u_k$. For every $n$,

$$F\left(\sum_{k=1}^{n} u_k\right) \le \sum_{k=1}^{n} F(u_k).$$

If, moreover, $F$ is continuous, it follows that

$$F\left(\sum_{k=1}^{\infty} u_k\right) \le \sum_{k=1}^{\infty} F(u_k) \le +\infty.$$

*Zabreiko's theorem* asserts that the converse is valid when $X$ is a Banach space.

**Theorem 3.2.10** *Let $X$ be a Banach space and let $F : X \to [0, +\infty[$ be a seminorm such that, for any convergent series $\sum\limits_{k=1}^{\infty} u_k$,*

$$F\left(\sum_{k=1}^{\infty} u_k\right) \le \sum_{k=1}^{\infty} F(u_k) \le +\infty.$$

*Then $F$ is continuous.*

**Proof** Let us define, for any $t > 0$, $G_t = \{u \in X : F(u) \le t\}$. Since $X = \bigcup\limits_{n=1}^{\infty} \overline{G_n}$, Baire's theorem implies the existence of $m$ such that $\overline{G_m}$ contains a closed ball $B[a, r]$. Using the propreties of $F$, we obtain

$$B[0, r] \subset \frac{1}{2}B[a, r] + \frac{1}{2}B[-a, r] \subset \overline{G}_{m/2} + \overline{G}_{m/2} \subset \overline{G}_m.$$

Let us define $t = m/r$, so that $B[0, 1]$ is contained in $\overline{G}_t$, and, for every $k$, $B[0, 1/2^k]$ is contained in $\overline{G}_{t/2^k}$. Let $u \in B[0, 1]$. There exists $u_1 \in G_t$ such that $\|u - u_1\| \le 1/2$. We construct by induction a sequence $(u_k)$ such that

$$u_k \in G_{t/2^{k-1}}, \quad \|u - u_1 - \ldots - u_k\| \le 1/2^k.$$

By assumption

$$F(u) = F\left(\sum_{k=1}^{\infty} u_k\right) \le \sum_{k=1}^{\infty} F(u_k) \le \sum_{k=1}^{\infty} t/2^{k-1} = 2t.$$

Since $u \in B[0, 1]$ is arbitrary, we obtain

$$\sup_{\substack{u \in X \\ \|u\|=1}} F(u) \le 2t.$$

It suffices then to use Proposition 3.2.9. □

Let $A$ be a linear mapping between two normed spaces $X$ and $Y$. If $A$ is continuous, then the graph of $A$ is closed in $X \times Y$:

$$u_n \xrightarrow{X} u, \; Au_n \xrightarrow{Y} v \quad \Rightarrow \quad v = Au.$$

The *closed graph theorem*, proven by S. Banach in 1932, asserts that the converse is valid when $X$ and $Y$ are Banach spaces.

**Theorem 3.2.11** *Let $X$ and $Y$ be Banach spaces and let $A: X \to Y$ be a linear mapping with a closed graph. Then $A$ is continuous.*

**Proof** Let us define on $X$ the seminorm $F(u) = \|Au\|$. Assume that the series $\sum\limits_{k=1}^{\infty} u_k$ converges to $u$ in $X$ and that $\sum\limits_{k=1}^{\infty} F(u_k) < +\infty$. Since $Y$ is a Banach space, $\sum\limits_{k=1}^{\infty} Au_k$ converges to $v$ in $Y$. But the graph of the linear mapping $A$ is closed, so that $v = Au$ and

$$F(u) = \|Au\| = \|v\| = \|\sum_{k=1}^{\infty} Au_k\| \le \sum_{k=1}^{\infty} \|Au_k\| = \sum_{k=1}^{\infty} F(u_k).$$

We conclude using Zabreiko's theorem:

$$\sup_{\substack{u \in X \\ \|u\|=1}} \|Au\| = \sup_{\substack{u \in X \\ \|u\|=1}} F(u) < +\infty.$$  □

The *open mapping theorem* was proved by J. Schauder in 1930.

**Theorem 3.2.12** *Let $X$ and $Y$ be Banach spaces and let $A \in \mathcal{L}(X, Y)$ be surjective. Then $\{Au : u \in X, \|u\| < 1\}$ is open in $Y$.*

**Proof** Let us define on $Y$ the seminorm $F(v) = \inf\{\|u\| : Au = v\}$. Assume that the series $\sum_{k=1}^{\infty} v_k$ converges to $v$ in $Y$ and that $\sum_{k=1}^{\infty} F(v_k) < +\infty$. Let $\varepsilon > 0$. For every $k$, there exists $u_k \in X$ such that

$$\|u_k\| \le F(v_k) + \varepsilon/2^k \quad \text{and} \quad Au_k = v_k.$$

Since $X$ is a Banach space, the series $\sum_{k=1}^{\infty} u_k$ converges to $u$ in $X$. Hence we obtain

$$\|u\| \le \sum_{k=1}^{\infty} \|u_k\| \le \sum_{k=1}^{\infty} F(v_k) + \varepsilon$$

and

$$Au = \sum_{k=1}^{\infty} Au_k = \sum_{k=1}^{\infty} v_k = v,$$

so that $F(v) \le \sum_{k=1}^{\infty} F(v_k) + \varepsilon$. Since $\varepsilon > 0$ is arbitrary, we conclude that $F(v) \le \sum_{k=1}^{\infty} F(v_k)$. Zabreiko's theorem implies that

$$\{Au : u \in X, \|u\| < 1\} = \{v \in Y : F(v) < 1\}$$

is open in $Y$.  □

## 3.3  Hilbert Spaces

Hilbert spaces are Banach spaces with a norm derived from a scalar product.

**Definition 3.3.1** A scalar product on the (real) vector space $X$ is a function

$$X \times X \to \mathbb{R} : (u, v) \mapsto (u|v)$$

such that

$(S_1)$  for every $u \in X \setminus \{0\}$, $(u|u) > 0$;
$(S_2)$  for every $u, v, w \in X$ and for every $\alpha, \beta \in \mathbb{R}$, $(\alpha u + \beta v|w) = \alpha(u|w) + \beta(v|w)$;
$(S_3)$  for every $u, v \in X$, $(u|v) = (v|u)$.

We define $||u|| = \sqrt{(u|u)}$. A (real) pre-Hilbert space is a (real) vector space together with a scalar product on that space.

**Proposition 3.3.2** *Let $u, v, w \in X$ and let $\alpha, \beta \in \mathbb{R}$. Then*

*(a)* $(u|\alpha v + \beta w) = \alpha(u|v) + \beta(u|w)$;
*(b)* $||\alpha u|| = |\alpha| \, ||u||$.

**Proposition 3.3.3** *Let $X$ be a pre-Hilbert space and let $u, v \in X$. Then*

*(a)* *(parallelogram identity)* $||u + v||^2 + ||u - v||^2 = 2||u||^2 + 2||v||^2$;
*(b)* *(polarization identity)* $(u|v) = \frac{1}{4}||u + v||^2 - \frac{1}{4}||u - v||^2$;
*(c)* *(Pythagorean identity)* $(u|v) = 0 \iff ||u + v||^2 = ||u||^2 + ||v||^2$.

***Proof*** Observe that

$$||u + v||^2 = ||u||^2 + 2(u|v) + ||v||^2, \tag{$*$}$$

$$||u - v||^2 = ||u||^2 - 2(u|v) + ||v||^2. \tag{$**$}$$

By adding and subtracting, we obtain parallelogram and polarization identities. The Pythagorean identity is clear.  □

**Proposition 3.3.4** *Let $X$ be a pre-Hilbert space and let $u, v \in X$. Then*

*(a)* *(Cauchy–Schwarz inequality)* $|(u|v)| \leq ||u|| \, ||v||$;
*(b)* *(Minkowski's inequality)* $||u + v|| \leq ||u|| + ||v||$.

***Proof*** It follows from $(*)$ and $(**)$ that for $||u|| = ||v|| = 1$,

$$|(u|v)| \leq \frac{1}{2}\Big(||u||^2 + ||v||^2\Big) = 1.$$

Hence for $u \neq 0 \neq v$, we obtain

$$\frac{|(u|v)|}{||u|| \, ||v||} = \left| \left( \frac{u}{||u||} \Big| \frac{v}{||v||} \right) \right| \leq 1.$$

By $(*)$ and the Cauchy–Schwarz inequality, we have

$$||u + v||^2 \leq ||u||^2 + 2||u|| \, ||v|| + ||v||^2 = \left( ||u|| + ||v|| \right)^2. \qquad \square$$

**Corollary 3.3.5** *(a) The function* $||u|| = \sqrt{(u|u)}$ *defines a norm on the pre-Hilbert space* $X$.
*(b) The function*

$$X \times X \to \mathbb{R} : (u, v) \mapsto (u|v)$$

*is continuous.*

**Definition 3.3.6** A family $(e_j)_{j \in J}$ in a pre-Hilbert space $X$ is orthonormal if

$$\begin{aligned} (e_j|e_k) &= 1, & j &= k, \\ &= 0, & j &\neq k. \end{aligned}$$

**Proposition 3.3.7 (Bessel's inequality)** *Let* $(e_n)$ *be an orthonormal sequence in a pre-Hilbert space* $X$ *and let* $u \in X$. *Then*

$$\sum_{n=0}^{\infty} |(u|e_n)|^2 \leq ||u||^2.$$

*Proof* It follows from the Pythagorean identity that

$$||u||^2 = \left\| u - \sum_{n=0}^{k} (u|e_n)e_n + \sum_{n=0}^{k} (u|e_n)e_n \right\|^2$$

$$= \left\| u - \sum_{n=0}^{k} (u|e_n)e_n \right\|^2 + \sum_{n=0}^{k} |(u|e_n)|^2$$

$$\geq \sum_{n=0}^{k} |(u|e_n)|^2. \qquad \square$$

**Proposition 3.3.8** *Let* $(e_0, \ldots, e_k)$ *be a finite orthonormal sequence in a pre-Hilbert space* $X$, $u \in X$, *and* $x_0, \ldots, x_k \in \mathbb{R}$. *Then*

$$\left\| u - \sum_{n=0}^{k} (u \mid e_n) e_n \right\| \leq \left\| u - \sum_{n=0}^{k} x_n e_n \right\|.$$

**Proof** It follows from the Pythagorean identity that

$$\left\| u - \sum_{n=0}^{k} x_n e_n \right\|^2 = \left\| u - \sum_{n=0}^{k} (u \mid e_n) e_n + \sum_{n=0}^{k} ((u \mid e_n) - x_n) e_n \right\|^2$$

$$= \left\| u - \sum_{n=0}^{k} (u \mid e_n) e_n \right\|^2 + \sum_{n=0}^{k} |(u \mid e_n) - x_n|^2. \qquad \square$$

**Definition 3.3.9** A Hilbert basis of a pre-Hilbert space $X$ is an orthonormal sequence generating a dense subspace of $X$.

**Proposition 3.3.10** *Let* $(e_n)$ *be a Hilbert basis of a pre-Hilbert space* $X$ *and let* $u \in X$. *Then*

*(a)* $u = \displaystyle\sum_{n=0}^{\infty} (u \mid e_n) e_n;$

*(b)* *(Parseval's identity)* $\|u\|^2 = \displaystyle\sum_{n=0}^{\infty} |(u \mid e_n)|^2.$

**Proof** Let $\varepsilon > 0$. By definition, there exists a sequence $x_0, \ldots, x_j \in \mathbb{R}$ such that

$$\left\| u - \sum_{n=0}^{j} x_n e_n \right\| < \varepsilon.$$

It follows from the preceding proposition that for $k \geq j$,

$$\left\| u - \sum_{n=0}^{k} (u \mid e_n) e_n \right\| < \varepsilon.$$

Hence $u = \displaystyle\sum_{n=0}^{\infty} (u \mid e_n) e_n$, and by Proposition 3.1.2,

$$\left\|\lim_{k\to\infty}\sum_{n=0}^{k}(u\mid e_n)e_n\right\|^2 = \lim_{k\to\infty}\left\|\sum_{n=0}^{k}(u\mid e_n)e_n\right\|^2 = \lim_{k\to\infty}\sum_{n=0}^{k}\left|(u\mid e_n)\right|^2 = \sum_{n=0}^{\infty}\left|(u\mid e_n)\right|^2.$$

$\square$

We characterize pre-Hilbert spaces having a Hilbert basis.

**Proposition 3.3.11** *Assume the existence of a sequence $(f_j)$ generating a dense subset of the normed space $X$. Then $X$ is separable.*

**Proof** By assumption, the space of (finite) linear combinations of $(f_j)$ is dense in $X$. Hence the space of (finite) linear combinations with rational coefficients of $(f_j)$ is dense in $X$. Since this space is countable, $X$ is separable. $\square$

**Proposition 3.3.12** *Let $X$ be an infinite-dimensional pre-Hilbert space. The following properties are equivalent:*

*(a) $X$ is separable;*
*(b) $X$ has a Hilbert basis.*

**Proof** By the preceding proposition, (b) implies (a).

If $X$ is separable, it contains a sequence $(f_j)$ generating a dense subspace. We may assume that $(f_j)$ is free. Since the dimension of $X$ is infinite, the sequence $(f_j)$ is infinite. We define by induction the sequences $(g_n)$ and $(e_n)$:

$$e_0 = f_0/\|f_0\|,$$

$$g_n = f_n - \sum_{j=0}^{n-1}(f_n\mid e_j)e_j, \ e_n = g_n/\|g_n\|, \quad n \geq 1.$$

The sequence $(e_n)$ generated from $(f_n)$ by the Gram–Schmidt orthonormalization process is a Hilbert basis of $X$. $\square$

**Definition 3.3.13** A Hilbert space is a complete pre-Hilbert space.

**Theorem 3.3.14 (Riesz–Fischer)** *Let $(e_n)$ be an orthonormal sequence in the Hilbert space $X$. The series $\sum_{n=0}^{\infty}c_n e_n$ converges if and only if $\sum_{n=0}^{\infty}c_n^2 < \infty$. Then*

$$\left\|\sum_{n=0}^{\infty}c_n e_n\right\|^2 = \sum_{n=0}^{\infty}c_n^2.$$

**Proof** Define $S_k = \sum_{n=0}^{k}c_n e_n$. The Pythagorean identity implies that for $j < k$,

$$||S_k - S_j||^2 = \left\| \sum_{n=j+1}^{k} c_n e_n \right\|^2 = \sum_{n=j+1}^{k} c_n^2.$$

Hence

$$\lim_{\substack{j \to \infty \\ j < k}} ||S_k - S_j||^2 = 0 \iff \lim_{\substack{j \to \infty \\ j < k}} \sum_{n=j+1}^{k} c_n^2 = 0 \iff \sum_{n=0}^{\infty} c_n^2 < \infty.$$

Since $X$ is complete, $(S_k)$ converges if and only if $\sum_{n=0}^{\infty} c_n^2 < \infty$. Then $\sum_{n=0}^{\infty} c_n e_n = \lim_{k \to \infty} S_k$, and by Proposition 3.1.2,

$$|| \lim_{k \to \infty} S_k ||^2 = \lim_{k \to \infty} ||S_k||^2 = \lim_{k \to \infty} \sum_{n=0}^{k} c_n^2 = \sum_{n=0}^{\infty} c_n^2. \qquad \square$$

*Examples* 1. Let $\mu$ be a positive measure on $\Omega$. We denote by $L^2(\Omega, \mu)$ the quotient of

$$\mathcal{L}^2(\Omega, \mu) = \left\{ u \in M(\Omega, \mu) : \int_{\Omega} |u|^2 d\mu < \infty \right\}$$

by the equivalence relation "equality almost everywhere." If $u, v \in L^2(\Omega, \mu)$, then $u + v \in L^2(\Omega, \mu)$. Indeed, almost everywhere on $\Omega$, we have

$$|u(x) + v(x)|^2 \leq 2(|u(x)|^2 + |v(x)|^2).$$

We define the scalar product

$$(u|v) = \int_{\Omega} uv \, d\mu$$

on the space $L^2(\Omega, \mu)$.

The scalar product is well defined, since almost everywhere on $\Omega$,

$$|u(x) \, v(x)| \leq \frac{1}{2}(|u(x)|^2 + |v(x)|^2).$$

By definition,

$$\|u\|_2 = \left( \int_\Omega |u|^2 d\mu \right)^{1/2}.$$

Convergence with respect to $\|.\|_2$ is convergence in quadratic mean. We will prove in Sect. 4.2, on Lebesgue spaces, that $L^2(\Omega, \mu)$ is a Hilbert space. If $\mu(\Omega) < \infty$, it follows from the Cauchy–Schwarz inequality that for every $u \in L^2(\Omega, \mu)$,

$$\|u\|_1 = \int_\Omega |u| \, d\mu \le \mu(\Omega)^{1/2} \|u\|_2.$$

Hence $L^2(\Omega, \mu) \subset L^1(\Omega, \mu)$, and the canonical injection is continuous.
2. Let $\Lambda_N$ be the Lebesgue measure on the open subset $\Omega$ of $\mathbb{R}^N$. We denote by $L^2(\Omega)$ the space $L^2(\Omega, \Lambda_N)$. Observe that

$$\frac{1}{x} \in L^2(]1, \infty[) \setminus L^1(]1, \infty[) \text{ and } \frac{1}{\sqrt{x}} \in L^1(]0, 1[) \setminus L^2(]0, 1[).$$

If $m(\Omega) < \infty$, the comparison theorem implies that for every $u \in \mathcal{BC}(\Omega)$,

$$\|u\|_2^2 = \int_\Omega u^2 dx \le m(\Omega) \|u\|_\infty^2.$$

Hence $\mathcal{BC}(\Omega) \subset L^2(\Omega)$, and the canonical injection is continuous.

**Theorem 3.3.15 (Vitali 1921, Dalzell 1945)** *Let $(e_n)$ be an orthonormal sequence in $L^2(]a, b[)$. The following properties are equivalent:*

*(a) $(e_n)$ is a Hilbert basis;*

*(b) for every $a \le t \le b$, $\displaystyle\sum_{n=1}^\infty \left( \int_a^t e_n(x) dx \right)^2 = t - a$;*

*(c) $\displaystyle\sum_{n=1}^\infty \int_a^b \left( \int_a^t e_n(x) dx \right)^2 dt = \frac{(b-a)^2}{2}$.*

**Proof** Property (b) follows from (a) and Parseval's identity applied to $\chi_{[a,t]}$. Property (c) follows from (b) and Levi's theorem. The converse is left to the reader. $\qquad\square$

*Example* The sequence $e_n(x) = \sqrt{\dfrac{2}{\pi}} \sin n \, x$ is orthonormal in $L^2(]0, \pi[)$. Since

$$\frac{2}{\pi} \sum_{n=1}^\infty \int_0^\pi \left( \int_0^t \sin n \, x \, dx \right)^2 dt = 3 \sum_{n=1}^\infty \frac{1}{n^2}$$

and since by a classical identity due to Euler,

$$\sum_{n=1}^{\infty} \frac{1}{n^2} = \frac{\pi^2}{6},$$

the sequence $(e_n)$ is a Hilbert basis of $L^2(]0, \pi[)$.

## 3.4 Spectral Theory

Spectral theory allows one to diagonalize symmetric compact operators.

**Definition 3.4.1** Let $X$ be a real vector space and let $A : X \to X$ be a linear mapping. The eigenvectors corresponding to the eigenvalue $\lambda \in \mathbb{R}$ are the nonzero solutions of

$$Au = \lambda u.$$

The multiplicity of $\lambda$ is the dimension of the space of solutions. The eigenvalue $\lambda$ is simple if its multiplicity is equal to 1. The rank of $A$ is the dimension of the range of $A$.

**Definition 3.4.2** Let $X$ be a pre-Hilbert space. A symmetric operator is a linear mapping $A : X \to X$ such that for every $u, v \in X$, $(Au|v) = (u|Av)$.

**Proposition 3.4.3** *Let $X$ be a pre-Hilbert space and $A : X \to X$ a symmetric continuous operator. Then*

$$||A|| = \sup_{\substack{u \in X \\ ||u|| = 1}} |(Au|u)|.$$

**Proof** It is clear that

$$a = \sup_{\substack{u \in X \\ ||u|| = 1}} |(Au|u)| \le b = \sup_{\substack{u, v \in X \\ ||u|| = ||v|| = 1}} |(Au|v)| = ||A||.$$

If $||u|| = ||v|| = 1$, it follows from the parallelogram identity that

$$|(Au|v)| = \frac{1}{4}|(A(u+v)|u+v) - (A(u-v)|u-v)|$$

$$\leq \frac{a}{4}[||u+v||^2 + ||u-v||^2]$$

$$= \frac{a}{4}[2||u||^2 + 2||v||^2] = a.$$

Hence $b = a$.                                                                     □

**Corollary 3.4.4**  *Under the assumptions of the preceding proposition, there exists a sequence $(u_n) \subset X$ such that*

$$||u_n|| = 1, ||Au_n - \lambda u_n|| \to 0, |\lambda_1| = ||A||.$$

**Proof**  Consider a maximizing sequence $(u_n)$:

$$||u_n|| = 1, |(Au_n|u_n)| \to \sup_{\substack{u \in X \\ ||u|| = 1}} |(Au|u)| = ||A||.$$

By passing if necessary to a subsequence, we can assume that $(Au_n|u_n) \to \lambda_1$, $|\lambda_1| = ||A||$. Hence

$$0 \leq ||Au_n - \lambda_1 u_n||^2 = ||Au_n||^2 - 2\lambda_1(Au_n|u_n) + \lambda_1^2||u_n||^2$$

$$\leq 2\lambda_1^2 - 2\lambda_1(Au_n|u_n) \to 0, \quad n \to \infty.$$                        □

**Definition 3.4.5**  Let $X$ and $Y$ be normed spaces. A mapping $A: X \to Y$ is compact if the set $\{Au: u \in X, ||u|| \leq 1\}$ is precompact in $Y$.

By Proposition 3.2.1, every linear compact mapping is continuous.

**Theorem 3.4.6**  *Let $X$ be a Hilbert space and let $A: X \to X$ be a symmetric compact operator. Then there exists an eigenvalue $\lambda_1$ of $A$ such that $|\lambda_1| = ||A||$.*

**Proof**  We can assume that $A \neq 0$. The preceding corollary implies the existence of a sequence $(u_n) \subset X$ such that

$$||u_n|| = 1, ||Au_n - \lambda_1 u_n|| \to 0, |\lambda_1| = ||A||.$$

Passing if necessary to a subsequence, we can assume that $Au_n \to v$. Hence $u_n \to u = \lambda_1^{-1}v$, $||u|| = 1$, and $Au = \lambda_1 u$.                                    □

**Theorem 3.4.7 (Poincaré's principle)** *Let $X$ be a Hilbert space and $A : X \to X$ a symmetric compact operator with infinite rank. Let there be given the eigenvectors $(e_1, \ldots, e_{n-1})$ and the corresponding eigenvalues $(\lambda_1, \ldots, \lambda_{n-1})$. Then there exists an eigenvalue $\lambda_n$ of $A$ such that*

$$|\lambda_n| = \max\{|(Au|u)| : u \in X, ||u|| = 1, (u|e_1) = \ldots = (u|e_{n-1}) = 0\}$$

*and $\lambda_n \to 0$, $n \to \infty$.*

***Proof*** The closed subspace of $X$

$$X_n = \{u \in X : (u|e_1) = \ldots = (u|e_{n-1}) = 0\}$$

is invariant by $A$. Indeed, if $u \in X_n$ and $1 \le j \le n - 1$, then

$$(Au|e_j) = (u|Ae_j) = \lambda_j(u|e_j) = 0.$$

Hence $A_n = A\Big|_{X_n}$ is a nonzero symmetric compact operator, and there exist an eigenvalue $\lambda_n$ of $A_n$ such that $|\lambda_n| = ||A_n||$ and a corresponding eigenvector $e_n \in X_n$ such that $||e_n|| = 1$. By construction, the sequence $(e_n)$ is orthonormal, and the sequence $(|\lambda_n|)$ is decreasing. Hence $|\lambda_n| \to d, n \to \infty$, and for $j \ne k$,

$$||Ae_j - Ae_k||^2 = \lambda_j^2 + \lambda_k^2 \to 2d^2, \quad j, k \to \infty.$$

Since $A$ is compact, $d = 0$.                                                                              $\square$

**Theorem 3.4.8** *Under the assumptions of the preceding theorem, for every $u \in X$, the series $\sum_{n=1}^{\infty}(u|e_n)e_n$ converges and $u - \sum_{n=1}^{\infty}(u|e_n)e_n$ belongs to the kernel of $A$:*

$$Au = \sum_{n=1}^{\infty}\lambda_n(u|e_n)e_n. \tag{*}$$

***Proof*** For every $k \ge 1$, $u - \sum_{n=1}^{k}(u|e_n)e_n \in X_{k+1}$. It follows from Proposition 3.3.8. that

$$\left\|Au - \sum_{n=1}^{k}\lambda_n(u|e_n)e_n\right\| \le ||A_{k+1}|| \left\|u - \sum_{n=1}^{k}(u|e_n)e_n\right\| \le ||A_{k+1}|| \, ||u|| \to 0, \; k \to \infty.$$

Bessel's inequality implies that $\sum_{n=1}^{\infty} |(u|e_n)|^2 \le ||u||^2$. We deduce from the Riesz–Fischer theorem that $\sum_{n=1}^{\infty} (u|e_n)e_n$ converges to $v \in X$. Since $A$ is continuous,

$$Av = \sum_{n=1}^{\infty} \lambda_n (u|e_n)e_n = Au$$

and $A(u - v) = 0$.                                                                                  □

Formula (∗) is the diagonalization of symmetric compact operators.

## 3.5  Comments

The de la Vallée Poussin criterion was proved in the beautiful paper [17].

The first proof of the Banach–Steinhaus theorem in Sect. 3.2 is due to Favard [22], and the second proof to Royden [66].

Theorem 3.2.10 is due to P.P. Zabreiko, *Funct. Anal. and Appl. 3 (1969) 70-72.*

Let us recall the elegant notion of vector space over the reals used by S. Banach in [6] :

Suppose that a non-empty set $E$ is given, and that to each ordered pair $(x, y)$ of elements of $E$ there corresponds an element $x + y$ of $E$ (called the *sum* of $x$ and $y$) and that for each number $t$ and $x \in E$ an element $tx$ of $E$ (called the *product* of the number $t$ with the element $x$) is defined in such a way that these operations, namely *addition* and *scalar multiplication* satisfy the following conditions (where $x$, $y$ and $z$ denote arbitrary elements of $E$ and $a, b$ are numbers):

1) $x + y = y + x$,
2) $x + (y + z) = (x + y) + z$,
3) $x + y = x + z$ *implies* $y = z$,
4) $a(x + y) = ax + ay$,
5) $(a + b)x = ax + bx$,
6) $a(bx) = (ab)x$,
7) $1 \cdot x = x$.

Under these hypotheses, we say that the set $E$ constitutes a *vector* or *linear* space. It is easy to see that there then exists exactly one element, which we denote by $\Theta$, such that $x + \Theta = x$ for all $x \in E$ and that the equality $ax = bx$ where $x \ne \Theta$ yields $a = b$; furthermore, that the equality $ax = ay$ where $a \ne 0$ implies $x = y$.
    Put, further, by definition :

$$-x = (-1)x \quad \text{and} \quad x - y = x + (-y).$$

The space $\mathcal{L}^1(\mathbb{R}^N)$ with the *pointwise sum*

$$(u + v)(x) = u(x) + v(x),$$

and the *scalar multiplication*

$$(a \cdot u)(x) = a \, u(x),$$

is *not* a vector space. Indeed one has in general to allow $-\infty$ and $+\infty$ as values of the elements of $\mathcal{L}^1(\mathbb{R}^N)$. Hence the pointwise sum and the scalar multiplication by 0 are not, in general, well defined. On the other hand the space $L^1(\Omega, \mu)$, with the pointwise sum and the scalar multiplication, is a vector space since it consists of equivalence classes of $\mu$-almost everywhere defined and finite function on $\Omega$.

## 3.6   Exercises for Chap. 3

1. Prove that $\mathcal{BC}(\Omega) \cap L^1(\Omega) \subset L^2(\Omega)$.
2. Define a sequence $(u_n) \subset \mathcal{BC}(]0, 1[)$ such that $||u_n||_1 \to 0$, $||u_n||_2 = 1$, and $||u_n||_\infty \to \infty$.
3. Define a sequence $(u_n) \subset \mathcal{BC}(\mathbb{R}) \cap L^1(\mathbb{R})$ such that $||u_n||_1 \to \infty$, $||u_n||_2 = 1$ and $||u_n||_\infty \to 0$.
4. Define a sequence $(u_n) \subset \mathcal{BC}(]0, 1[)$ converging simply to $u$ such that $||u_n||_\infty = ||u||_\infty = ||u_n - u||_\infty = 1$.
5. Define a sequence $(u_n) \subset L^1(]0, 1[)$ such that $||u_n||_1 \to 0$ and for every $0 < x < 1$, $\varlimsup_{n \to \infty} u_n(x) = 1$. *Hint*: Use characteristic functions of intervals.
6. On the space $C([0, 1])$ with the norm $||u||_1 = \displaystyle\int_0^1 |u(x)|dx$, is the linear functional

$$f : C([0, 1]) \to \mathbb{R} : u \mapsto u(1/2)$$

continuous?
7. Let $X$ be a normed space such that every normally convergent series converges. Prove that $X$ is a Banach space.
8. A linear functional defined on a normed space is continuous if and only if its kernel is closed. If this is not the case, the kernel is dense.
9. Is it possible to derive the norm on $L^1(]0, 1[)$ (respectively $\mathcal{BC}(]0, 1[)$) from a scalar product?
10. Prove *Lagrange's identity* in pre-Hilbert spaces:

$$\big|\big| ||v||u - ||u||v \big|\big|^2 = 2||u||^2||v||^2 - 2||u|| \, ||v||(u|v).$$

11. Let $X$ be a pre-Hilbert space and $u, v \in X \setminus \{0\}$. Then

$$\left|\left| \frac{u}{||u||^2} - \frac{v}{||v||^2} \right|\right| = \frac{||u - v||}{||u|| \, ||v||}.$$

Let $f, g, h \in X$. Prove *Ptolemy's inequality*:

$$||f|| \, ||g - h|| \leq ||h|| \, ||f - g|| + ||g|| \, ||h - f||.$$

12. (The Jordan–von Neumann theorem.) Assume that the parallelogram identity is valid in the normed space $X$. Then it is possible to derive the norm from a scalar product. Define

$$(u|v) = \frac{1}{4}\left(||u + v||^2 - ||u - v||^2\right).$$

Verify that

$$(f + g|h) + (f - g|h) = 2(f|h),$$

$$(u|h) + (v|h) = 2\left(\frac{u + v}{2}\Big|h\right) = (u + v|h).$$

13. Let $f$ be a linear functional on $L^2(]0, 1[)$ such that $u \geq 0 \Rightarrow \langle f, u \rangle \geq 0$. Prove, by contradiction, that $f$ is continuous with respect to the norm $||.||_2$. Prove that $f$ is not necessarily continuous with respect to the norm $||.||_1$.

14. Prove that every symmetric operator defined on a Hilbert space is continuous. *Hint*: If this were not the case, there would exist a sequence $(u_n)$ such that $||u_n|| = 1$ and $||Au_n|| \to \infty$. Then use the Banach–Steinhaus theorem to obtain a contradiction.

15. In a Banach space an algebraic basis is either finite or uncountable. *Hint*: Use Baire's theorem.

16. Assume that $\mu(\Omega) < \infty$. Let $(u_n) \subset L^1(\Omega, \mu)$ be such that

  (a) $\displaystyle\sup_n \int_\Omega |u_n|\ell n(1 + |u_n|)d\mu < +\infty;$

  (b) $(u_n)$ converges almost everywhere to $u$.

  Then $u_n \to u$ in $L^1(\Omega, \mu)$.

17. Let us define, for $n \geq 1$, $u_n(x) = \dfrac{\cos 3^n x}{n}$.

  (a) The series $\displaystyle\sum_{n=1}^{\infty} u_n$ converges in $L^2(]0, 2\pi[)$.

  (b) For every $x \in A = \{2k\pi/3^j : j \in \mathbb{N}, k \in \mathbb{Z}\}$, $\displaystyle\sum_{n=1}^{\infty} u_n(x) = +\infty$.

  (c) For every $x \in B = \{(2k + 1)\pi/3^j : j \in \mathbb{N}, k \in \mathbb{Z}\}$, $\displaystyle\sum_{n=1}^{\infty} u_n(x) = -\infty$.

  (d) The sets $A$ and $B$ are dense in $\mathbb{R}$.

# Chapter 4
# Lebesgue Spaces

## 4.1 Convexity

The notion of convexity plays a basic role in functional analysis and in the theory of inequalities.

**Definition 4.1.1** A subset $C$ of a vector space $X$ is convex if for every $u, v \in C$ and every $0 < \lambda < 1$, we have $(1 - \lambda)x + \lambda y \in C$.

A point $x$ of the convex set $C$ is internal if for every $y \in X$, there exists $\varepsilon > 0$ such that $x + \varepsilon y \in C$. The set of internal points of $C$ is denoted by int $C$.

A subset $C$ of $X$ is a cone if for every $x \in C$ and every $\lambda > 0$, we have $\lambda x \in C$.

Let $C$ be a convex set. A function $F : C \to ] - \infty, +\infty]$ is convex if for every $x, y \in C$ and every $0 < \lambda < 1$, we have $F((1 - \lambda)x + \lambda y) \leq (1 - \lambda)F(x) + \lambda F(y)$.

A function $F : C \to [-\infty, +\infty[$ is concave if $-F$ is convex.

Let $C$ be a cone. A function $F : C \to ] - \infty, +\infty]$ is positively homogeneous if for every $x \in C$ and every $\lambda > 0$, we have $F(\lambda x) = \lambda F(x)$.

*Examples* Every linear function is convex, concave, and positively homogeneous. Every norm is convex and positively homogeneous. Open balls and closed balls in a normed space are convex.

**Proposition 4.1.2** *The upper envelope of a family of convex (respectively positively homogeneous) functions is convex (respectively positively homogeneous).*

**Lemma 4.1.3** *Let $Y$ be a hyperplane of a real vector space $X$, $f : Y \to \mathbb{R}$ linear and $F : X \to ] - \infty, +\infty]$ convex and positively homogeneous such that $f \leq F$ on $Y$ and*

$$Y \cap \text{int}\{x \in X : F(x) < \infty\} \neq \phi.$$

© Springer Nature Switzerland AG 2022
M. Willem, *Functional Analysis*, Cornerstones,
https://doi.org/10.1007/978-3-031-09149-0_4

*Then there exists $g : X \to \mathbb{R}$ linear such that $g \leq F$ on $X$ and $g\big|_Y = f$.*

**Proof** There exists $z \in X$ such that $X = Y \oplus \mathbb{R}z$. We must prove the existence of $c \in \mathbb{R}$ such that for every $y \in Y$ and every $t \in \mathbb{R}$,

$$\langle f, y \rangle + ct \leq F(y + tz).$$

Since $F$ is positively homogeneous, it suffices to verify that for every $u, v \in Y$,

$$\langle f, u \rangle - F(u - z) \leq c \leq F(v + z) - \langle f, v \rangle.$$

For every $u, v \in Y$, we have by assumption that

$$\langle f, u \rangle + \langle f, v \rangle \leq F(u + v) \leq F(u - z) + F(v + z).$$

We define

$$a = \sup_{u \in Y} \langle f, u \rangle - F(u - z) \leq b = \inf_{v \in Y} F(v + z) - \langle f, v \rangle.$$

Let $u \in Y \cap \text{int}\{x \in X : F(x) < \infty\}$. For $t$ large enough, $F(tu - z) = tF(u - z/t) < +\infty$. Hence $-\infty < a$. Similarly, $b < +\infty$. We can choose any $c \in [a, b]$.  □

Let us state a cornerstone of functional analysis, the *Hahn–Banach theorem*.

**Theorem 4.1.4** *Let $Y$ be a subspace of a separable normed space $X$, and let $f \in \mathcal{L}(Y, \mathbb{R})$. Then there exists $g \in \mathcal{L}(X, \mathbb{R})$ such that $||g|| = ||f||$ and $g\big|_Y = f$.*

**Proof** Let $(z_n)$ be a sequence dense in $X$. We define $f_0 = f$, $Y_0 = Y$, and $Y_n = Y_{n-1} + \mathbb{R}z_n$, $n \geq 1$. Let there be $f_n \in \mathcal{L}(Y_n, \mathbb{R})$ such that $||f_n|| = ||f||$ and $f_n\big|_{Y_{n-1}} = f_{n-1}$. If $Y_{n+1} = Y_n$, we define $f_{n+1} = f_n$. If this is not the case, the preceding lemma implies the existence of $f_{n+1} : Y_{n+1} \to \mathbb{R}$ linear such that $f_{n+1}\big|_{Y_n} = f_n$ and for every $x \in Y_{n+1}$,

$$\langle f_{n+1}, x \rangle \leq ||f|| \, ||x||.$$

On $Z = \bigcup_{n=0}^{\infty} Y_n$ we define $h$ by $h\big|_{Y_n} = f_n$, $n \geq 0$. The space $Z$ is dense in $X$, $h \in \mathcal{L}(Z, \mathbb{R})$, $||h|| = ||f||$, and $h\big|_Y = f$. Finally, by Proposition 3.2.3, there exists $g \in \mathcal{L}(X, \mathbb{R})$ such that $||g|| = ||h||$ and $g\big|_Z = h$.  □

*Notation* The *dual* of a normed space $X$ is defined by $X^* = \mathcal{L}(X, \mathbb{R})$. Let us recall that the norm on $X^*$ is defined by

$$||g|| = \sup_{\substack{u \in X \\ ||u|| \leq 1}} |\langle g, u \rangle| = \sup_{\substack{u \in X \\ ||u|| \leq 1}} \langle g, u \rangle.$$

**Theorem 4.1.5** *Let $Z$ be a subspace of a separable normed space $X$, and let $u \in X \setminus \overline{Z}$. Then*

$$0 < d(u, Z) = \max\{\langle g, u \rangle : g \in X^*, ||g|| \leq 1, g\big|_Z = 0\}.$$

*In particular if $u \in X \setminus \{0\}$, then*

$$||u|| = \max_{\substack{g \in X^* \\ ||g|| \leq 1}} \langle g, u \rangle = \max_{\substack{g \in X^* \\ ||g|| \leq 1}} |\langle g, u \rangle|.$$

*Proof* Let us first prove that

$$c = \sup \left\{ \langle g, u \rangle : g \in X^* : ||g|| \leq 1, g\big|_Z = 0 \right\} \leq \delta = d(u, Z).$$

Assume that $||g|| \leq 1$ and $g\big|_Z = 0$. Then, for every $z \in Z$,

$$\langle g, u \rangle = \langle g, u - z \rangle \leq ||g|| \, ||u - z|| \leq ||u - z||,$$

so that $\langle g, u \rangle \leq \delta$ and $c \leq \delta$.

It suffices then to prove the existence of $g \in X^*$ such that $||g|| \leq 1$, $g\big|_Z = 0$ and $\langle g, u \rangle = \delta$. Let us define the functional $f$ on $Y = \mathbb{R}u \oplus Z$ by

$$\langle f, tu + z \rangle = t\delta.$$

Since, for $t \neq 0$,

$$\langle f, tu + z \rangle \leq |t|\delta \leq |t| \, ||u + z/t|| = ||tu + z||,$$

the functional $f$ is such that $||f|| \leq 1$. The preceding theorem implies the existence of $g \in X^*$ such that $||g|| = ||f|| \leq 1$ and $g\big|_Y = f$. In particular $\langle g, u \rangle = \delta$ and $g\big|_Z = 0$. □

The next theorem is due to P. Roselli and the author. Let us define

$$C_+ = \{(x_1, x_2) \in \mathbb{R}^2 : x_1 \geq 0, x_2 \geq 0\}.$$

**Theorem 4.1.6 (Convexity Inequality)** *Let* $F : C_+ \to \mathbb{R}$ *be a positively homogeneous function, and let* $u_j \in L^1(\Omega, \mu)$ *be such that* $u_j \geq 0$, $\int_\Omega u_j d\mu > 0$, $j = 1, 2$. *If* $F$ *is convex, then*

$$F\left(\int_\Omega u_1 d\mu, \int_\Omega u_2 d\mu\right) \leq \int_\Omega F(u_1, u_2) d\mu.$$

*If* $F$ *is concave, the reverse inequality holds.*

**Proof** We define $F(x) = +\infty$, $x \in \mathbb{R}^2 \setminus C_+$, and $y_j = \int_\Omega u_j d\mu$, $j = 1, 2$. Lemma 4.1.3 implies the existence of $\alpha, \beta \in \mathbb{R}$ such that

$$F(y_1, y_2) = \alpha y_1 + \beta y_2 \text{ and, for all } x_1, x_2 \in \mathbb{R}, \alpha x_1 + \beta x_2 \leq F(x_1, x_2). \quad (*)$$

For every $0 \leq \lambda \leq 1$, we have

$$\alpha(1 - \lambda) + \beta\lambda \leq F(1 - \lambda, \lambda) \leq (1 - \lambda)F(1, 0) + \lambda F(0, 1),$$

so that $c = \sup_{0 \leq \lambda \leq 1} |F(1 - \lambda, \lambda)| < \infty$. Since

$$\left|F(u_1, u_2)\right| \leq c(u_1 + u_2),$$

the comparison theorem implies that $F(u_1, u_2) \in L^1(\Omega, \mu)$. We conclude from $(*)$ that

$$F\left(\int_\Omega u_1 d\mu, \int_\Omega u_2 d\mu\right) = \alpha \int_\Omega u_1 d\mu + \beta \int_\Omega u_2 d\mu$$

$$= \int_\Omega \alpha u_1 + \beta u_2 d\mu$$

$$\leq \int_\Omega F(u_1, u_2) d\mu. \qquad \square$$

**Lemma 4.1.7** *Let* $F : C_+ \to \mathbb{R}$ *be a continuous and positively homogeneous function. If* $F(., 1)$ *is convex (respectively concave), then* $F$ *is convex (respectively concave).*

**Proof** Assume that $F(., 1)$ is convex. It suffices to prove that for every $x, y \in \overset{\circ}{C}_+$, $F(x + y) \leq F(x) + F(y)$. The preceding inequality is equivalent to

$$F\left(\frac{x_1 + y_1}{x_2 + y_2}, 1\right) \leq \frac{x_2}{x_2 + y_2} F\left(\frac{x_1}{x_2}, 1\right) + \frac{y_2}{x_2 + y_2} F\left(\frac{y_1}{y_2}, 1\right). \qquad \square$$

*Remark* Define $F$ on $\mathbb{R}^2$ by

$$F(y, z) = -\sqrt{yz}, \quad (y, z) \in C_+,$$
$$= +\infty, \quad (y, z) \in \mathbb{R}^2 \setminus C_+.$$

The function $F$ is positively homogeneous and, by the preceding lemma, is convex on $C_+$, hence on $\mathbb{R}^2$. It is clear that $0 = F$ on $Y = \mathbb{R} \times \{0\}$. There is no linear function $g : \mathbb{R}^2 \to \mathbb{R}$ such that $g \leq F$ on $\mathbb{R}^2$ and $g = 0$ on $Y$.

The convexity inequality implies a version of the Cauchy–Schwarz inequality: if $v, w \in L^1(\Omega, \mu)$, then

$$\int_\Omega |vw|^{1/2} d\mu \leq \left( \int_\Omega |v| d\mu \right)^{1/2} \left( \int_\Omega |w| d\mu \right)^{1/2}.$$

**Definition 4.1.8** Let $1 < p < \infty$. The exponent $p'$ conjugate to $p$ is defined by $1/p + 1/p' = 1$. On the Lebesgue space

$$\mathcal{L}^p(\Omega, \mu) = \left\{ u \in M(\Omega, \mu) : \int_\Omega |u|^p d\mu < \infty \right\},$$

we define the functional $||u||_p = \left( \int_\Omega |u|^p d\mu \right)^{1/p}$.

**Theorem 4.1.9** *Let* $1 < p < \infty$.

*(a) (Hölder's inequality.) Let $v \in \mathcal{L}^p(\Omega, \mu)$ and $w \in \mathcal{L}^{p'}(\Omega, \mu)$. Then*

$$\int_\Omega |vw| d\mu \leq ||v||_p ||w||_{p'}.$$

*(b) (Minkowski's inequality.) Let $v, w \in \mathcal{L}^p(\Omega, \mu)$. Then*

$$||v + w||_p \leq ||v||_p + ||w||_p.$$

*(c) (Hanner's inequalities.) Let $v, w \in \mathcal{L}^p(\Omega, \mu)$. If $2 \leq p < \infty$, then*

$$||v + w||_p^p + ||v - w||_p^p \leq (||v||_p + ||w||_p)^p + \big|||v||_p - ||w||_p\big|^p.$$

*If $1 < p \leq 2$, the reverse inequality holds.*

***Proof*** On $C_+$, we define the continuous positively homogeneous functions

$$F(x_1, x_2) = x_1^{1/p} x_2^{1/p'},$$

$$G(x_1, x_2) = (x_1^{1/p} + x_2^{1/p})^p,$$

$$H(x_1, x_2) = (x_1^{1/p} + x_2^{1/p})^p + |x_1^{1/p} - x_2^{1/p}|^p.$$

Inequality (a) follows from the convexity inequality applied to $F$ and $u = (|v|^p, |w|^{p'})$. Inequality (b) follows from the convexity inequality applied to $G$ and $u = (|v|^p, |w|^p)$. Finally, inequalities (c) follow from the convexity inequality applied to $H$ and $u = (|v|^p, |w|^p)$. When $v = 0$ or $w = 0$, the inequalities are obvious.

On $[0, +\infty[$, we define $f = F(., 1)$, $g = G(., 1)$, $h = H(., 1)$. It is easy to verify that

$$f''(x) = \frac{1-p}{p^2} x^{\frac{1}{p}-2},$$

$$g''(x) = \frac{1-p}{p} x^{-\frac{1}{p}-1}(x^{-\frac{1}{p}} + 1)^{p-2},$$

$$h''(x) = \frac{1-p}{p} x^{-\frac{1}{p}-1} \left[ (x^{-\frac{1}{p}} + 1)^{p-2} - |x^{-\frac{1}{p}} - 1|^{p-2} \right].$$

Hence $f$ and $g$ are concave. If $2 \leq p < \infty$, then $h$ is concave, and if $1 < p \leq 2$, then $h$ is convex. It suffices then to use the preceding lemma.                    □

## 4.2  Lebesgue Spaces

Let $\mu : \mathcal{L} \to \mathbb{R}$ be a positive measure on the set $\Omega$.

**Definition 4.2.1**  Let $1 \leq p < \infty$. The space $L^p(\Omega, \mu)$ is the quotient of $\mathcal{L}^p(\Omega, \mu)$ by the equivalence relation "equality almost everywhere." By definition,

$$||u||_{L^p(\Omega, \mu)} = ||u||_p = \left( \int_\Omega |u|^p d\mu \right)^{1/p}.$$

When $\Lambda_N$ is the Lebesgue measure on the open subset $\Omega$ of $\mathbb{R}^N$, the space $L^p(\Omega, \Lambda_N)$ is denoted by $L^p(\Omega)$.

In practice, we identify the elements of $L^p(\Omega, \mu)$ and the functions of $\mathcal{L}^p(\Omega, \mu)$.

**Proposition 4.2.2**  *Let $1 \leq p < \infty$. Then the space $L^p(\Omega, \mu)$ with the norm $||.||_p$ is a normed space.*

**Proof**  Minkowski's inequality implies that if $u, v \in L^p(\Omega, \mu)$, then $u + v \in L^p(\Omega, \mu)$ and

$$||u + v||_p \leq ||u||_p + ||v||_p.$$

It is clear that if $u \in L^p(\Omega, \mu)$ and $\lambda \in \mathbb{R}$, then $\lambda u \in L^p(\Omega, \mu)$ and $||\lambda u||_p = |\lambda| \, ||u||_p$. Finally, if $||u||_p = 0$, then $u = 0$ almost everywhere and $u = 0$ in $L^p(\Omega, \mu)$. □

The next inequalities follow from Hölder's inequality.

**Proposition 4.2.3 (Generalized Hölder's Inequality)** *Let* $1 < p_j < \infty$, $u_j \in L^{p_j}(\Omega, \mu)$, $1 \leq j \leq k$, *and* $1/p_1 + \ldots + 1/p_k = 1$. *Then* $\prod_{j=1}^{k} u_j \in L^1(\Omega, \mu)$ *and*

$$\int_{\Omega} \prod_{j=1}^{k} |u_j| d\mu \leq \prod_{j=1}^{k} ||u_j||_{p_j}.$$

**Proposition 4.2.4 (Interpolation Inequality)** *Let* $1 \leq p < q < r < \infty$,

$$\frac{1}{q} = \frac{1 - \lambda}{p} + \frac{\lambda}{r},$$

*and* $u \in L^p(\Omega, \mu) \cap L^r(\Omega, \mu)$. *Then* $u \in L^q(\Omega, \mu)$ *and*

$$||u||_q \leq ||u||_p^{1-\lambda} ||u||_r^{\lambda}.$$

**Proposition 4.2.5** *Let* $1 \leq p < q < \infty$, $\mu(\Omega) < \infty$, *and* $u \in L^q(\Omega, \mu)$. *Then* $u \in L^p(\Omega, \mu)$ *and*

$$||u||_p \leq \mu(\Omega)^{\frac{1}{p} - \frac{1}{q}} ||u||_q.$$

**Proposition 4.2.6** *Let* $1 \leq p < \infty$ *and* $(u_n) \subset L^p(\Omega, \mu)$ *be such that*

*(a)* $||u_n||_p \to ||u||_p$, $n \to \infty$;
*(b)* $u_n$ *converges to* $u$ *almost everywhere.*

*Then* $||u_n - u||_p \to 0$, $n \to \infty$.

**Proof** Since almost everywhere

$$0 \leq 2^p(|u_n|^p + |u|^p) - |u_n - u|^p,$$

Fatou's lemma ensures that

$$2^{p+1} \int_\Omega |u|^p d\mu \leq \varliminf \int_\Omega \left[ 2^p (|u_n|^p + |u|^p) - |u_n - u|^p \right] d\mu$$

$$= 2^{p+1} \int |u|^p d\mu - \varlimsup \int_\Omega |u_n - u|^p d\mu.$$

Hence $\varlimsup \|u_n - u\|_p^p \leq 0$. $\qquad\qquad\qquad\qquad\qquad\qquad\qquad\qquad\qquad\qquad\square$

The next result is more precise.

**Theorem 4.2.7 (Brezis–Lieb Lemma)** *Let* $1 \leq p < \infty$ *and let* $(u_n) \subset L^p(\Omega, \mu)$ *be such that*

*(a)* $c = \sup_n \|u_n\|_p < \infty$;

*(b)* $u_n$ *converges to* $u$ *almost everywhere.*

*Then* $u \in L^p(\Omega, \mu)$ *and*

$$\lim_{n \to \infty} \left( \|u_n\|_p^p - \|u_n - u\|_p^p \right) = \|u\|_p^p.$$

**Proof** By Fatou's lemma, $\|u\|_p \leq c$. Let $\varepsilon > 0$. There exists, by homogeneity, $c(\varepsilon) > 0$ such that for every $a, b \in \mathbb{R}$,

$$\left| |a + b|^p - |a|^p - |b|^p \right| \leq \varepsilon |a|^p + c(\varepsilon) |b|^p.$$

We deduce from Fatou's lemma that

$$\int_\Omega c(\varepsilon) |u|^p d\mu \leq \varliminf_{n \to \infty} \int_\Omega \varepsilon |u_n - u|^p + c(\varepsilon) |u|^p - \left| |u_n|^p - |u_n - u|^p - |u|^p \right| d\mu$$

$$\leq (2c)^p \varepsilon + \int_\Omega c(\varepsilon) |u|^p d\mu - \varlimsup_{n \to \infty} \int_\Omega \left| |u_n|^p - |u_n - u|^p - |u|^p \right| d\mu,$$

or

$$\varlimsup_{n \to \infty} \int_\Omega \left| |u_n|^p - |u_n - u|^p - |u|^p \right| d\mu \leq (2c)^p \varepsilon.$$

Since $\varepsilon > 0$ is arbitrary, the proof is complete. $\qquad\qquad\qquad\qquad\qquad\square$

We define

$$\begin{aligned} R_h(s) &= s + h, \quad && s \leq -h, \\ &= 0, \quad && |s| < h, \\ &= s - h, \quad && s \geq h. \end{aligned}$$

**Theorem 4.2.8 (Degiovanni–Magrone)** *Let* $\mu(\Omega) < \infty$, $1 \leq p < \infty$, *and* $(u_n) \subset L^p(\Omega, \mu)$ *be such that* *(a)*  $c = \sup_n ||u_n||_p < \infty;$

*(b)  $u_n$ converges to u almost everywhere.*

*Then*

$$\lim_{n \to \infty} \left( ||u_n||_p^p - ||R_h u_n||_p^p \right) = ||u||_p^p - ||R_h u||_p^p.$$

**Proof** Let us define

$$f(s) = |s|^p - |R_h(s)|^p.$$

For every $\varepsilon > 0$, there exists $c(\varepsilon) > 0$ such that

$$|f(s) - f(t)| \leq \varepsilon \big| |s|^p + |t|^p \big| + c(\varepsilon).$$

It follows from Fatou's lemma that

$$2\varepsilon \int_\Omega |u|^p d\mu + c(\varepsilon) m(\Omega) \leq \varliminf_{n \to \infty} \int_\Omega \varepsilon \left( |u_n|^p + |u|^p \right) + c(\varepsilon) - \big| f(u_n) - f(u) \big| d\mu$$

$$\leq \varepsilon \, c^p + \varepsilon \int_\Omega |u|^p d\mu + c(\varepsilon) \mu(\Omega) - \varlimsup_{n \to \infty} \int_\Omega \big| f(u_n) - f(u) \big| d\mu.$$

Hence

$$\varlimsup_{n \to \infty} \int_\Omega \big| f(u_n) - f(u) \big| d\mu \leq \varepsilon \, c^p.$$

Since $\varepsilon > 0$ is arbitrary, the proof is complete.                                        □

**Theorem 4.2.9 (F. Riesz, 1910)** *Let $1 \leq p < \infty$. Then the space $L^p(\Omega, \mu)$ is complete.*

**Proof** Let $(u_n)$ be a Cauchy sequence in $L^p(\Omega, \mu)$. There exists a subsequence $v_j = u_{n_j}$ such that for every $j$,

$$||v_{j+1} - v_j||_p \leq 1/2^j.$$

We define the sequence

$$f_k = \sum_{j=1}^k |v_{j+1} - v_j|.$$

Minkowski's inequality ensures that

$$\int_{\Omega} f_k^p d\mu \le \left(\sum_{j=1}^{k} 1/2^j\right)^p < 1.$$

Levi's theorem implies the almost everywhere convergence of $f_k$ to $f \in L^p(\Omega, \mu)$. Hence $v_k$ converges almost everywhere to a function $u$. For $m \ge k + 1$, it follows from Minkowski's inequality that

$$\int_{\Omega} |v_m - v_k|^p d\mu \le \left(\sum_{j=k}^{m-1} 1/2^j\right)^p \le (2/2^k)^p.$$

By Fatou's lemma, we obtain

$$\int_{\Omega} |u - v_k|^p d\mu \le (2/2^k)^p.$$

In particular, $u = u - v_1 + v_1 \in L^p(\Omega, \mu)$. We conclude by invoking the Cauchy condition:

$$\|u - u_k\|_p \le \|u - v_k\|_p + \|v_k - u_k\|_p \le 2/2^k$$
$$+\|u_{n_k} - u_k\|_p \to 0, \quad k \to \infty. \qquad \square$$

**Proposition 4.2.10** *Let $1 \le p < \infty$ and let $u_n \to u$ in $L^p(\Omega, \mu)$. Then there exist subsequences $v_j = u_{n_j}$ and $g \in L^p(\Omega, \mu)$ such that almost everywhere,*

$$|v_j| \le g \text{ and } v_j \to u, \quad j \to \infty.$$

***Proof*** If the sequence $(u_n)$ converges in $L^p(\Omega, \mu)$, it satisfies the Cauchy condition by Proposition 1.2.3. The subsequence $(v_j)$ in the proof of the preceding theorem converges almost everywhere to $u$, and for every $j$,

$$|v_j| \le |v_1| + \sum_{j=1}^{\infty} |v_{j+1} - v_j| = |v_1| + f \in L^p(\Omega, \mu). \qquad \square$$

**Theorem 4.2.11 (Density Theorem)** *Let $1 \le p < \infty$ and $\mathcal{L} \subset L^p(\Omega, \mu)$. Then $\mathcal{L}$ is dense in $L^p(\Omega, \mu)$.*

***Proof*** Let $u \in L^p(\Omega, \mu)$. Since $u$ is measurable with respect to $\mu$ on $\Omega$, there exists a sequence $(u_n) \subset \mathcal{L}$ such that $u_n \to u$ almost everywhere. We define

$$v_n = \max(\min(|u_n|, u), -|u_n|).$$

By definition, $|v_n| \leq |u_n|$, and almost everywhere,

$$|v_n - u|^p \leq |u|^p \in L^1, |v_n - u|^p \to 0, \quad n \to \infty.$$

It follows from Lebesgue's dominated convergence theorem that $||v_n - u||_p \to 0$, $n \to \infty$. Hence

$$Y = \{u \in L^p(\Omega, \mu) : \text{there exists } f \in \mathcal{L} \text{ such that } |u| \leq f \text{ almost everywhere}\}$$

is dense in $L^p(\Omega, \mu)$. It suffices to prove that $\mathcal{L}$ is dense in $Y$.

Let $u \in Y$, $f \in \mathcal{L}$ be such that $|u| \leq f$ almost everywhere and $(u_n) \subset \mathcal{L}$ such that $u_n \to u$ almost everywhere. We define

$$w_n = \max(\min(f, u_n), -f).$$

By definition, $w_n \in \mathcal{L}$ and, almost everywhere,

$$|w_n - u|^p \leq 2^p f^p \in L^1, |w_n - u|^p \to 0, \quad n \to \infty.$$

It follows from Lebesgue's dominated convergence theorem that $||w_n - u||_p \to 0$, $n \to \infty$. Hence $\mathcal{L}$ is dense in $Y$. $\qquad \square$

**Theorem 4.2.12** *Let $\Omega$ be open in $\mathbb{R}^N$ and $1 \leq p < \infty$. Then the space $L^p(\Omega)$ is separable.*

**Proof** By the preceding theorem, $\mathcal{K}(\Omega)$ is dense in $L^p(\Omega)$. Proposition 2.3.2 implies that for every $u \in \mathcal{K}(\Omega)$,

$$u_j = \sum_{k \in \mathbb{Z}^N} u(k/2^j) f_{j,k}$$

converges to $u$ in $L^p(\Omega)$. We conclude the proof using Proposition 3.3.11. $\qquad \square$

## 4.3 Regularization

*La logique parfois engendre des monstres. Depuis un demi-siècle on a vu surgir une foule de fonctions bizarres qui semblent s'efforcer de ressembler aussi peu que possible aux honnêtes fonctions qui servent à quelque chose.*

Henri Poincaré

Regularization by convolution allows one to approximate locally integrable functions by infinitely differentiable functions.

**Definition 4.3.1** Let $\Omega$ be an open subset of $\mathbb{R}^N$. The space of test functions on $\Omega$ is defined by

$$\mathcal{D}(\Omega) = \{u \in C^\infty(\mathbb{R}^N) : \text{spt } u \text{ is a compact subset of } \Omega\}.$$

Let $\alpha = (\alpha_1, \ldots, \alpha_N) \in \mathbb{N}^N$ be a multi-index. By definition,

$$|\alpha| = \alpha_1 + \ldots + \alpha_N, \quad D^\alpha = \partial_1^{\alpha_1} \ldots \partial_N^{\alpha_N}, \quad \partial_j = \frac{\partial}{\partial x_j}.$$

Using a function defined by Cauchy in 1821, we shall verify that 0 is not the only element in $\mathcal{D}(\Omega)$.

**Proposition 4.3.2** *The function defined on $\mathbb{R}$ by*

$$f(x) = \exp(1/x), \quad x < 0,$$
$$= 0, \qquad\qquad x \geq 0,$$

*is infinitely differentiable.*

**Proof** Let us prove by induction that for every $n$ and every $x < 0$,

$$f^{(n)}(0) = 0, \quad f^{(n)}(x) = P_n(1/x) \exp(1/x),$$

where $P_n$ is a polynomial. The statement is true for $n = 0$. Assume that it is true for $n$. We obtain

$$\lim_{x \to 0^-} \frac{f^{(n)}(x) - f^n(0)}{x} = \lim_{x \to 0^-} \frac{P_n(1/x) \exp(1/x)}{x} = 0.$$

Hence $f^{(n+1)}(0) = 0$. Finally, we have for $x < 0$,

$$f^{(n+1)}(x) = (-1/x^2)(P_n(1/x) + P_n'(1/x)) \exp(1/x) = P_{n+1}(1/x) \exp(1/x). \quad \square$$

**Definition 4.3.3** We define on $\mathbb{R}^N$ the function

$$\rho(x) = c^{-1} \exp(1/(|x|^2 - 1)), \quad |x| < 1,$$
$$= 0, \qquad\qquad\qquad\qquad |x| \geq 1,$$

where

$$c = \int_{B(0,1)} \exp(1/(|x|^2 - 1))dx.$$

The regularizing sequence $\rho_n(x) = n^N \rho(nx)$ is such that

$$\rho_n \in \mathcal{D}(\mathbb{R}^N), \quad \mathrm{spt}\,\rho_n = B[0, 1/n], \quad \int_{\mathbb{R}^N} \rho_n\,dx = 1, \quad \rho_n \geq 0.$$

**Definition 4.3.4** Let $\Omega$ be an open set of $\mathbb{R}^N$. By definition, $\omega \subset\subset \Omega$ if $\omega$ is open and $\overline{\omega}$ is a compact subset of $\Omega$. We define, for $1 \leq p < \infty$,

$$L^p_{\mathrm{loc}}(\Omega) = \{u : \Omega \to \mathbb{R} : \text{for all } \omega \subset\subset \Omega, u\big|_{\omega} \in L^p(\omega)\}.$$

A sequence $(u_n)$ converges to $u$ in $L^p_{\mathrm{loc}}(\Omega)$ if for every $\omega \subset\subset \Omega$,

$$\int_{\omega} |u_n - u|^p dx \to 0, \quad n \to \infty.$$

**Definition 4.3.5** Let $u \in L^1_{\mathrm{loc}}(\Omega)$ and $v \in \mathcal{K}(\mathbb{R}^N)$ be such that $\mathrm{spt}\,v \subset B[0, 1/n]$. For $n \geq 1$, the convolution $v * u$ is defined on

$$\Omega_n = \{x \in \Omega : d(x, \partial\Omega) > 1/n\}$$

by

$$v * u(x) = \int_{\Omega} v(x - y)u(y)dy = \int_{B(0,1/n)} v(y)u(x - y)dy.$$

If $|y| < 1/n$, the translation of $u$ by $y$ is defined on $\Omega_n$ by $\tau_y u(x) = u(x - y)$.

**Proposition 4.3.6** *Let* $u \in L^1_{\mathrm{loc}}(\Omega)$ *and* $v \in \mathcal{D}(\mathbb{R}^N)$ *be such that* $\mathrm{spt}\,v \subset B[0, 1/n]$. *Then* $v * u \in C^\infty(\Omega_n)$, *and for every* $\alpha \in \mathbb{N}^N$, $D^\alpha(v * u) = (D^\alpha v) * u$.

**Proof** Let $|\alpha| = 1$ and $x \in \Omega_n$. There exists $r > 0$ such that $B[x, r] \subset \Omega_n$. Hence

$$\omega = B(x, r + 1/n) \subset\subset \Omega,$$

and for $0 < |\varepsilon| < r$,

$$\frac{v * u(x + \varepsilon\alpha) - v * u(x)}{\varepsilon} = \int_{\omega} \frac{v(x + \varepsilon\alpha - y) - v(x - y)}{\varepsilon} u(y)dy.$$

But

$$\lim_{\substack{\varepsilon \to 0 \\ \varepsilon \neq 0}} \frac{v(x + \varepsilon\alpha - y) - v(x - y)}{\varepsilon} = D^\alpha v(x - y)$$

and

$$\sup_{\substack{y \in \omega \\ 0 < |\varepsilon| < r}} \left| \frac{v(x + \varepsilon\alpha - y) - v(x - y)}{\varepsilon} \right| < \infty.$$

Lebesgue's dominated convergence theorem implies that

$$D^\alpha(v * u)(x) = \int_\omega D^\alpha v(x - y)u(y)dy = (D^\alpha v) * u(x).$$

It is easy to conclude the proof by induction.                                        □

**Lemma 4.3.7** *Let $\omega \subset\subset \Omega$.*

*(a) Let $u \in C(\Omega)$. Then for every $n$ large enough,*

$$\sup_{x \in \omega} |\rho_n * u(x) - u(x)| \leq \sup_{|y| < 1/n} \sup_{x \in \omega} |\tau_y u(x) - u(x)|.$$

*(b) Let $u \in L^p_{\text{loc}}(\Omega)$, $1 \leq p < \infty$. Then for every $n$ large enough,*

$$\|\rho_n * u - u\|_{L^p(\omega)} \leq \sup_{|y| < 1/n} \|\tau_y u - u\|_{L^p(\omega)}.$$

***Proof*** For every $n$ large enough, $\omega \subset\subset \Omega_n$. Let $u \in C(\Omega)$. Since

$$\int_{B(0,1/n)} \rho_n(y)dy = 1,$$

we obtain for every $x \in \omega$,

$$\left| \rho_n * u(x) - u(x) \right| = \left| \int_{B(0,1/n)} \rho_n(y)\Big(u(x - y) - u(x)\Big)dy \right|$$

$$\leq \sup_{|y| < 1/n} \sup_{x \in \omega} \left| u(x - y) - u(x) \right|.$$

Let $u \in L^p_{\text{loc}}(\Omega)$, $1 \leq p < \infty$. By Hölder's inequality, for every $x \in \omega$, we have

$$\left| \rho_n * u(x) - u(x) \right| = \left| \int_{B(0,1/n)} \rho_n(y)\Big(u(x-y) - u(x)\Big)dy \right|$$

$$\leq \left( \int_{B(0,1/n)} \rho_n(y)\big|u(x-y) - u(x)\big|^p dy \right)^{1/p}.$$

Fubini's theorem implies that

$$\int_\omega \big|\rho_n * u(x) - u(x)\big|^p dx \leq \int_\omega dx \int_{B(0,1/n)} \rho_n(y)\big|u(x-y) - u(x)\big|^p dy$$

$$= \int_{B(0,1/n)} dy \int_\omega \rho_n(y)\big|u(x-y) - u(x)\big|^p dx$$

$$\leq \sup_{|y|<1/n} \int_\omega \big|u(x-y) - u(x)\big|^p dx. \qquad \square$$

**Lemma 4.3.8 (Continuity of Translations)** *Let* $\omega \subset\subset \Omega$.

(a) *Let* $u \in C(\Omega)$. *Then* $\lim_{y\to 0} \sup_{x\in\omega} |\tau_y u(x) - u(x)| = 0$.

(b) *Let* $u \in L^p_{\mathrm{loc}}(\Omega)$, $1 \leq p < \infty$. *Then* $\lim_{y\to 0} ||\tau_y u - u||_{L^p(\omega)} = 0$.

**Proof** We choose an open subset $U$ such that $\omega \subset\subset U \subset\subset \Omega$. If $u \in C(\Omega)$, then property (a) follows from the uniform continuity of $u$ on $U$.

Let $u \in L^p_{\mathrm{loc}}(\Omega)$, $1 \leq p < \infty$, and $\varepsilon > 0$. The density theorem implies the existence of $v \in \mathcal{K}(U)$ such that $||u - v||_{L^p(U)} \leq \varepsilon$. By (a), there exists $0 < \delta < d(\omega, \partial U)$ such that for every $|y| < \delta$, $\sup_{x\in\omega}|\tau_y v(x) - v(x)| \leq \varepsilon$. We obtain for every $|y| < \delta$,

$$||\tau_y u - u||_{L^p(\omega)} \leq ||\tau_y u - \tau_y v||_{L^p(\omega)} + ||\tau_y v - v||_{L^p(\omega)} + ||v - u||_{L^p(\omega)}$$

$$\leq 2||u - v||_{L^p(U)} + m(\omega)^{1/p} \sup_{x\in\omega}|\tau_y v(x) - v(x)|$$

$$\leq (2 + m(\omega)^{1/p})\varepsilon.$$

Since $\varepsilon > 0$ is arbitrary, the proof is complete. $\qquad \square$

We deduce from the preceding lemmas the following *regularization theorem*.

**Theorem 4.3.9**

(a) *Let $u \in C(\Omega)$. Then $\rho_n * u$ converges uniformly to $u$ on every compact subset of $\Omega$.*

(b) *Let $u \in L^p_{\text{loc}}(\Omega)$, $1 \le p < \infty$. Then $\rho_n * u$ converges to $u$ in $L^p_{\text{loc}}(\Omega)$.*

The following consequences are fundamental.

**Theorem 4.3.10 (Annulation Theorem)** *Let $u \in L^1_{\text{loc}}(\Omega)$ be such that for every $v \in \mathcal{D}(\Omega)$,*

$$\int_\Omega v(x)u(x)dx = 0.$$

*Then $u = 0$ almost everywhere on $\Omega$.*

**Proof** By assumption, for every $n$, $\rho_n * u = 0$ on $\Omega_n$.                             □

**Theorem 4.3.11** *Let $1 \le p < \infty$. Then $\mathcal{D}(\Omega)$ is dense in $L^p(\Omega)$.*

**Proof** By the density theorem, $\mathcal{K}(\Omega)$ is dense in $L^p(\Omega)$. Let $u \in \mathcal{K}(\Omega)$. There exists an open set $\omega$ such that spt $u \subset \omega \subset\subset \Omega$. For $j$ large enough, the support of $u_j = \rho_j * u$ is contained in $\omega$. Since $u_j \in C^\infty(\mathbb{R}^N)$ by Proposition 4.3.6, $u_j \in \mathcal{D}(\Omega)$. The regularization theorem ensures that $u_j \to u$ in $L^p(\Omega)$.                             □

**Definition 4.3.12** A partition of unity subordinate to the covering of the compact subset $\Gamma$ of $\mathbb{R}^N$ by the open sets $U_1, \ldots, U_k$ is a sequence $\psi_1, \ldots, \psi_k$ such that

(a) $\psi_j \in \mathcal{D}(U_j)$, $\psi_j \ge 0$, $j = 1, \ldots, k$;

(b) $\displaystyle\sum_{j=1}^k \psi_j = 1$ on $\Gamma$, $\displaystyle\sum_{j=1}^k \psi_j \le 1$ on $\mathbb{R}^N$.

Let us prove the *theorem of partition of unity*.

**Theorem 4.3.13** *Let $U_1, \ldots, U_k$ be a covering by open sets of the compact subset $\Gamma$ of $\mathbb{R}^N$. Then there exists a partition of unity subordinates to $U_1, \ldots, U_k$.*

**Proof** Let $K$ be a compact subset of the open subset $U$ of $\mathbb{R}^N$. We choose an open set $\omega$ such that $K \subset \omega \subset\subset U$. For $n$ large enough, $\varphi = \rho_n * \chi_\omega$ is such that $\varphi \in \mathcal{D}(U)$, $\varphi = 1$ on $K$ and $0 \le \varphi \le 1$ on $\mathbb{R}^N$.

For $n$ large enough, the finite sequence

$$F_j = \{x : d(x, \mathbb{R}^N \setminus U_j) \ge 1/n\}, \quad j = 1, \ldots, k$$

is a covering of $\Gamma$ by closed sets. Indeed if this is not the case, there exists, by the compactness of $\Gamma$, $x \subset \Gamma \setminus \displaystyle\bigcup_{j=1}^k U_j$. This is a contradiction.

By the first part of the proof, there exists, for $j = 1, \ldots, k$, $\varphi_j \in \mathcal{D}(U_j)$ such that $\varphi_j = 1$ on $\Gamma \cap F_j$ and $0 \leq \varphi_j \leq 1$ on $\mathbb{R}^N$. Let us define the functions

$$\psi_1 = \varphi_1,$$
$$\psi_2 = \varphi_2(1 - \varphi_1),$$
$$\ldots$$
$$\psi_k = \varphi_k(1 - \varphi_1) \ldots (1 - \varphi_{k-1}).$$

It is easy to prove, by a finite induction, that

$$\psi_1 + \ldots + \psi_k = 1 - (1 - \varphi_1) \ldots (1 - \varphi_k).$$

Assume that $x \in \Gamma$. There exists $j$ such that $x \in F_j$. By definition, we conclude that $\varphi_j(x) = 1$ and $\psi_1(x) + \ldots + \psi_k(x) = 1$. □

Now we consider Euclidean space.

**Proposition 4.3.14** *Let* $1 \leq p < \infty$ *and* $u \in L^p(\mathbb{R}^N)$. *Then* $||\rho_n * u||_p \leq ||u||_p$ *and* $\rho_n * u \to u$ *in* $L^p(\mathbb{R}^N)$.

*Proof* It follows from Hölder's inequality that

$$\left| \rho_n * u(x) \right| = \left| \int_{\mathbb{R}^N} u(y) \rho_n(x - y) dy \right| \leq \left| \int_{\mathbb{R}^N} |u(y)|^p \rho_n(x - y) dy \right|^{1/p}.$$

Fubini's theorem implies that

$$\int_{\mathbb{R}^N} \left| \rho_n * u(x) \right|^p dx \leq \int_{\mathbb{R}^N} dx \int_{\mathbb{R}^N} |u(y)|^p \rho_n(x - y) dy$$

$$= \int_{\mathbb{R}^N} dy \int_{\mathbb{R}^N} |u(y)|^p \rho_n(x - y) dx$$

$$= \int_{\mathbb{R}^N} |u(y)|^p dy.$$

Hence $||\rho_n * u||_p \leq ||u||_p$.

Let $u \in L^p(\mathbb{R}^N)$ and $\varepsilon > 0$. The density theorem implies the existence of $v \in \mathcal{K}(\mathbb{R}^N)$ such that $||u - v||_p \leq \varepsilon$. By the regularization theorem, $\rho_n * v \to v$ in $L^p(\mathbb{R}^N)$. Hence there exists $m$ such that for every $n \geq m$, $||\rho_n * v - v||_p \leq \varepsilon$. We obtain for every $n \geq m$ that

$$||\rho_n * u - u||_p \leq ||\rho_n * (u - v)||_p + ||\rho_n * v - v||_p + ||v - u||_p \leq 3\varepsilon.$$

Since $\varepsilon > 0$ is arbitrary, the proof is complete. □

**Proposition 4.3.15** *Let* $1 \leq p < \infty$, $f \in L^p(\mathbb{R}^N)$, *and* $g \in \mathcal{K}(\mathbb{R}^N)$. *Then*

$$\int_{\mathbb{R}^N} (\rho_n * f)g \, dx = \int_{\mathbb{R}^N} f(\rho_n * g) dx.$$

***Proof*** Fubini's theorem and the parity of $\rho$ imply that

$$\begin{aligned}
\int_{\mathbb{R}^N} (\rho_n * f)(x)g(x)dx &= \int_{\mathbb{R}^N} dx \int_{\mathbb{R}^N} \rho_n(x - y)f(y)g(x)dy \\
&= \int_{\mathbb{R}^N} dy \int_{\mathbb{R}^N} \rho_n(x - y)f(y)g(x)dx \\
&= \int_{\mathbb{R}^N} (\rho_n * g)(y)f(y)dy.
\end{aligned}$$
□

## 4.4  Compactness

We prove a variant of *Ascoli's theorem*.

**Theorem 4.4.1** *Let* $X$ *be a precompact metric space, and let* $S$ *be a set of uniformly continuous functions on* $X$ *such that*

(a) $c = \sup\limits_{u \in S} \sup\limits_{x \in X} |u(x)| < \infty$;

(b) *for every* $\varepsilon > 0$, *there exists* $\delta > 0$ *such that* $\sup\limits_{u \in S} \omega_u(\delta) \leq \varepsilon$.

*Then* $S$ *is precompact in* $\mathcal{BC}(X)$.

***Proof*** Let $\varepsilon > 0$ and let $\delta$ corresponds to $\varepsilon$ by (b). There exists a finite covering of the precompact space $X$ by balls $B[x_1, \delta], \ldots, B[x_k, \delta]$. There exists also a finite covering of $[-c, c]$ by intervals $[y_1 - \varepsilon, y_1 + \varepsilon], \ldots, [y_n - \varepsilon, y_n + \varepsilon]$. Let us denote by $J$ the (finite) set of mappings from $\{1, \ldots, k\}$ to $\{1, \ldots, n\}$. For every $j \in J$, we define

$$S_j = \{u \in S : |u(x_1) - y_{j(1)}| \leq \varepsilon, \ldots, |u(x_k) - y_{j(k)}| \leq \varepsilon\}.$$

By definition, $(S_j)_{j \in J}$ is a covering of $S$. Let $u, v \in S_j$ and $x \in X$. There exists $m$ such that $d(x, x_m) \leq \delta$. We have

$$|u(x_m) - y_{j(m)}| \leq \varepsilon, \quad |v(x_m) - y_{j(m)}| \leq \varepsilon$$

and, by (b),

$$|u(x) - u(x_m)| \leq \varepsilon, \quad |v(x) - v(x_m)| \leq \varepsilon.$$

Hence $|u(x) - v(x)| \leq 4\varepsilon$, and since $x \in X$ is arbitrary, $||u - v||_\infty \leq 4\varepsilon$. If $S_j$ is nonempty, then $S_j \subset B[u, 4\varepsilon]$. Since $\varepsilon > 0$ is arbitrary, $S$ is precompact in $\mathcal{BC}(X)$ by Fréchet's criterion. $\qquad \square$

We prove a variant of *M. Riesz's theorem* (1933).

**Theorem 4.4.2** *Let $\Omega$ be an open subset of $\mathbb{R}^N$, $1 \leq p < \infty$, and let $S \subset L^p(\Omega)$ be such that*

(a) $c = \sup\limits_{u \in S} ||u||_{L^p(\Omega)} < \infty$;

(b) *for every $\varepsilon > 0$, there exists $\omega \subset\subset \Omega$ such that* $\sup\limits_{u \in S} \int_{\Omega \setminus \omega} |u|^p dx \leq \varepsilon^p$;

(c) *for every $\omega \subset\subset \Omega$,* $\lim\limits_{y \to 0} \sup\limits_{u \in S} ||\tau_y u - u||_{L^p(\omega)} = 0$.

*Then $S$ is precompact in $L^p(\Omega)$.*

**Proof** Let $\varepsilon > 0$ and let $\omega$ corresponds to $\varepsilon$ by (b). Assumption (c) implies the existence of $0 < \delta < d(\omega, \partial\Omega)$ such that for every $|y| \leq \delta$,

$$\sup_{u \in S} ||\tau_y u - u||_{L^p(\omega)} \leq \varepsilon.$$

We choose $n > 1/\delta$. We deduce from Lemma 4.3.7 that

$$\sup_{u \in S} ||\rho_n * u - u||_{L^p(\omega)} \leq \sup_{u \in S} \sup_{|y| < 1/n} ||\tau_y u - u||_{L^p(\omega)} \leq \varepsilon. \qquad (*)$$

We define

$$U = \{x \in \mathbb{R}^N : d(x, \omega) < 1/n\} \subset\subset \Omega.$$

Let us prove that the family $R = \{\rho_n * u|_\omega : u \in S\}$ satisfies the assumptions of Ascoli's theorem in $\mathcal{BC}(\omega)$.

1. By (a), for every $u \in S$ and for every $x \in \omega$, we have

$$|\rho_n * u(x)| \leq \int_U \rho_n(x - z)|u(z)|dz \leq \sup_{\mathbb{R}^N} |\rho_n| \, ||u||_{L^1(U)} \leq c_1.$$

2. By (a), for every $u \in S$ and for every $x, y \in \omega$, we have

$$|\rho_n * u(x) - \rho_n * u(y)| \leq \int_U |\rho_n(x - z) - \rho_n(y - z)| \, |u(z)|dz$$

$$\leq \sup_z |\rho_n(x - z) - \rho_n(y - z)| \, ||u||_{L^1(U)} \leq c_2|x - y|.$$

Hence $R$ is precompact in $\mathcal{BC}(\omega)$. Since

$$||v||_{L^p(\omega)} \leq m(\omega)^{1/p} \sup_{\omega} |v|,$$

$R$ is precompact in $L^p(\omega)$. But then $(*)$ implies the existence of a finite covering of $S|_\omega$ in $L^p(\omega)$ by balls of radius $2\varepsilon$. Assumption (b) ensures the existence of a finite covering of $S$ in $L^p(\Omega)$ by balls of radius $3\varepsilon$. Since $\varepsilon > 0$ is arbitrary, $S$ is precompact in $L^p(\Omega)$ by Fréchet's criterion.                                                  □

## 4.5  Comments

Figure 4.1 gives a geometric interpretation of Lemma 4.1.3. It is contained in the *Lectures on Analysis* by G. Choquet (W.A. Benjamin, New York, 1969).

Proofs of the Hahn–Banach theorem without the axiom of choice (in separable spaces) are given in the treatise by Garnir et al. [28] and in the lectures by Favard [22].

The convexity inequality is due to Roselli and the author [64]. In contrast to Jensen's inequality [36], it is not restricted to probability measures. But we have to consider positively homogeneous functions. See [16] for the relations between convexity and lower semicontinuity.

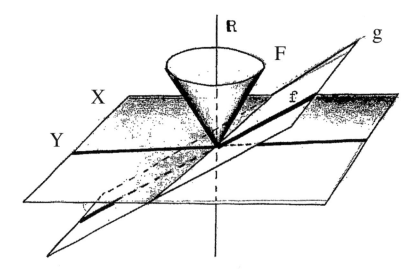

**Fig. 4.1** Lemma of the Hahn-Banach theorem

## 4.6   Exercises for Chap. 4

1. (Young's inequality.) Let $1 < p < \infty$. Then for every $a, b \geq 0$,

$$ab \leq \frac{a^p}{p} + \frac{b^{p'}}{p'}.$$

First proof: $A = \ell n\, a^p$, $B = \ell n\, b^{p'}$, $\exp\left(\dfrac{A}{p} + \dfrac{B}{p'}\right) \leq \dfrac{\exp A}{p} + \dfrac{\exp B}{p'}$.

Second proof: $\dfrac{b^{p'}}{p'} = \sup\limits_{a \geq 0}\left(ab - \dfrac{a^p}{p}\right)$.

2. (Hölder's inequality.) Let $1 < p < \infty$. If $||u||_p \neq 0 \neq ||v||_{p'}$, then by Young's inequality,

$$\int_\Omega \left| \frac{u}{||u||_p} \frac{v}{||v||_{p'}} \right| d\mu \leq 1.$$

3. (Minkowski's inequality.) Prove that

   (a) $||u||_p = \sup\limits_{||w||_{p'}=1} \int_\Omega uw\, d\mu$

   (b) $||u + v||_p \leq ||u||_p + ||v||_p$

4. (Minkowski's inequality.) Let $1 < p < \infty$ and define, on $L^p(\Omega, \mu)$, the convex function $G(u) = \int_\Omega |u|^p d\mu$. Then with $\lambda = ||v||_p/(||u||_p + ||v||_p)$,

$$G\left(\frac{u+v}{||u||_p + ||v||_p}\right) = G\left((1-\lambda)\frac{u}{||u||_p} + \lambda\frac{v}{||v||_p}\right)$$

$$\leq (1-\lambda)G\left(\frac{u}{||u||_p}\right) + \lambda G\left(\frac{v}{||v||_p}\right) = 1.$$

   Hence $||u + v||_p \leq ||u||_p + ||v||_p$.

5. (Jensen's inequality)

   (a) Let $f : [0, +\infty[ \to \mathbb{R}$ be a convex function and $y > 0$. There exists $\alpha, \beta \in \mathbb{R}$ such that

$$f(y) = \alpha y + \beta \text{ and, for all } x \geq 0, \alpha x + \beta \leq f(x).$$

   (b) Let $f : [0, +\infty[ \to \mathbb{R}$ be a convex function. Let $\mu$ be a positive measure on $\Omega$ such that $\mu(\Omega) = 1$, and let $u \in L^1(\Omega, \mu)$ be such that $u \geq 0$ and $\int_\Omega u\, d\mu > 0$. Then

$$f\left(\int_\Omega u\, d\mu\right) \le \int_\Omega f(u)d\mu \le +\infty.$$

If $f$ is concave, the reverse inequality holds.

6. Assume that $\mu(\Omega) = 1$. Then for every $u \in L^1(\Omega, \mu)$, $u \ge 0$,

$$0 \le \exp\int_\Omega \ell n\, u\, d\mu \le \int_\Omega u\, d\mu \le \ell n \int_\Omega \exp u\, d\mu \le +\infty.$$

7. Let $\Omega = B(0, 1) \subset \mathbb{R}^N$. Then

$$\lambda p + N > 0 \Longleftrightarrow |x|^\lambda \in L^p(\Omega), \lambda p + N < 0 \Longleftrightarrow |x|^\lambda \in L^p(\mathbb{R}^N \setminus \overline{\Omega}).$$

8. A differentiable function $u : \mathbb{R} \to \mathbb{R}$ satisfies

$$x^2 u'(x) + u(x) = 0$$

if and only if $u(x) = cf(x)$, where $c \in \mathbb{R}$ and $f$ is the function defined in Proposition 4.3.2.

9. Let $1 < p < \infty$, $(u_n) \subset L^1(\Omega, \mu)$ and let $u: \Omega \to \mathbb{R}$ be $\mu$-measurable. Then the following properties are equivalent:

  (a) $\|u_n - u\|_p \to 0$, $n \to \infty$;
  (b) $(u_n)$ converges in measure to $u$ and $\{|u_n|^p : n \in \mathbb{N}\}$ is equi-integrable.

10. (Rising sun lemma, F. Riesz, 1932.) Let $g \in C([a, b])$. The set

$$E = \left\{a < x < b : g(x) < \max_{[x,b]} g\right\}$$

consists of a finite or countable union of disjoint intervals $]a_k, b_k[$ such that $g(a_k) \le g(b_k)$. *Hint*: If $a_k < x < b_k$, then $g(x) < g(b_k)$.

11. (Maximal inequality, Hardy–Littlewood, 1930.) Let $u \in L^1(]a, b[)$, $u \ge 0$. The *maximal function* defined on $]a, b[$ by

$$Mu(x) = \sup_{x<y<b} \frac{1}{y-x} \int_x^y u(s)ds$$

satisfies, for every $t > 0$,

$$|\{Mu > t\}| \le t^{-1} \int_a^b u(s)ds.$$

*Hint*: Use the rising sun lemma with

$$g(x) = \int_a^x u(s)ds - tx.$$

12. (Lebesgue's differentiability theorem) Let $u \in L^1(]a, b[)$. Prove that for almost every $a < x < b$,

$$\lim_{\substack{y \to x \\ y > x}} \frac{1}{y - x} \int_x^y |u(s) - u(x)|ds = 0.$$

*Hint*: Use Theorem 4.3.11 and the maximal inequality.

13. (Godunova's inequality) Let $f: [0, +\infty[ \to [0, +\infty[$ be convex, and let $u: \mathbb{R} \to [0, +\infty[$ be Lebesgue-measurable. Then

$$\int_0^\infty f\left(\int_0^x u(t)\frac{dt}{x}\right)\frac{dx}{x} \leq \int_0^\infty f(u(x))\frac{dx}{x} \leq +\infty.$$

Hint:

$$\int_0^\infty f\left(\int_0^x u(t)\frac{dt}{x}\right)\frac{dx}{x} \leq \int_0^\infty dx \int_0^x f(u(t))\frac{dt}{x^2}$$

$$= \int_0^\infty dt \int_t^\infty f(u(t))\frac{dx}{x^2}$$

$$= \int_0^\infty f(u(t))\frac{dt}{t}.$$

14. (Hardy's inequality) Let $1 < p < \infty$ and $v: \mathbb{R} \to [0, +\infty[$ be Lebesgue-measurable. Then

$$\int_0^\infty \left[\int_0^x v(t)\frac{dt}{x}\right]^p dx \leq \left(\frac{p}{p-1}\right)^p \int_0^\infty v^p(x)dx \leq +\infty. \qquad (*)$$

Verify that this inequality is optimal using the family

$$f_\varepsilon(x) = 1, \qquad 0 < x \leq 1,$$

$$= x^{-\varepsilon-1/p}, \qquad x > 1.$$

Hint. Godunova's inequality

$$\int_0^\infty \left[\int_0^x u(t)\frac{dt}{x}\right]^p \frac{dx}{x} \leq \int_0^\infty u^p(x)\frac{dx}{x}$$

is equivalent to $(*)$ where

$$v(x) = x^{-1/p} u(x^{1-1/p}).$$

15. (Knopp's inequality) Let $v \colon \mathbb{R} \to [0, +\infty[$ be Lebesgue-measurable. Then

$$\int_0^\infty \exp\left(\int_0^x v(t)\frac{dt}{x}\right) dx \le e \int_0^\infty \exp v(x)dx \le +\infty. \qquad (**)$$

Hint. Godunova's inequality

$$\int_0^\infty \exp\left(\int_0^x u(t)\frac{dt}{x}\right)\frac{dx}{x} \le \int_0^\infty \exp u(x)\frac{dx}{x}$$

is equivalent to $(**)$ where

$$v(x) = u(x) - \ln x.$$

# Chapter 5
# Duality

## 5.1  Weak Convergence

A fruitful process in functional analysis is to associate to every normed space $X$ the dual space $X^*$ of linear continuous functionals on $X$.

**Definition 5.1.1** Let $X$ be a normed space. The dual $X^*$ of $X$ is the space of linear continuous functionals on $X$. A sequence $(f_n) \subset X^*$ converges weakly to $f \in X^*$ if $(f_n)$ converges simply to $f$. We then write $f_n \rightharpoonup f$.

Let us translate Proposition 3.2.5 and Corollary 3.2.7.

**Proposition 5.1.2** *Let $Z$ be a dense subset of a normed space $X$ and $(f_n) \subset X^*$ such that:*

*(a)* $\sup_{n} ||f_n|| < \infty$;
*(b)* *for every $v \in Z$, $\langle f_n, v \rangle$ converges.*

*Then $(f_n)$ converges weakly to $f \in X^*$ and*

$$||f|| \leq \lim_{n \to \infty} ||f_n||.$$

**Theorem 5.1.3 (Banach–Steinhaus)** *Let $X$ be a Banach space and $(f_n) \subset X^*$ simply convergent to $f$. Then $(f_n)$ is bounded, $f \in X^*$, and*

$$||f|| \leq \lim_{n \to \infty} ||f_n||.$$

**Theorem 5.1.4 (Banach)** *Let $X$ be a separable normed space. Then every sequence bounded in $X^*$ contains a weakly convergent subsequence.*

© Springer Nature Switzerland AG 2022
M. Willem, *Functional Analysis*, Cornerstones,
https://doi.org/10.1007/978-3-031-09149-0_5

**Proof** A Cantor diagonal argument will be used. Let $(f_n)$ be bounded in $X^*$, and let $(v_k)$ be dense in $X$. Since $(\langle f_n, v_1 \rangle)$ is bounded, there exists a subsequence $(f_{1,n})$ of $(f_n)$ such that $\langle f_{1,n}, v_1 \rangle$ converges as $n \to \infty$. By induction, for every $k$, there exists a subsequence $(f_{k,n})$ of $(f_{k-1,n})$ such that $\langle f_{k,n}, v_k \rangle$ converges as $n \to \infty$. The sequence $g_n = f_{n,n}$ is bounded, and for every $k$, $\langle g_n, v_k \rangle$ converges as $n \to \infty$. By Proposition 5.1.2, $(g_n)$ converges weakly in $X^*$.                                                                        □

*Example (Weak Convergence)* Let us prove that $\mathcal{BC}(]0, 1[)$ is not separable. We define on this space the functionals $\langle f_n, u \rangle = u(1/n)$. It is clear that $||f_n|| = 1$. For every strictly increasing sequence $(n_k)$, there exists $u \in \mathcal{BC}(]0, 1[)$ such that $u(1/n_k) = (-1)^k$. Hence,

$$\lim_{k \to \infty} \langle f_{n_k}, u \rangle = -1, \quad \overline{\lim_{k \to \infty}} \langle f_{n_k}, u \rangle = 1,$$

and the sequence $(f_{n_k})$ is not weakly convergent.

Let $\Omega$ be an open subset of $\mathbb{R}^N$. We define

$$\mathcal{K}_+(\Omega) = \{u \in \mathcal{K}(\Omega) : \text{for all } x \in \Omega, u(x) \geq 0\}.$$

**Theorem 5.1.5** *Let* $\mu : \mathcal{K}(\Omega) \to \mathbb{R}$ *be a linear functional such that for every* $u \in \mathcal{K}_+(\Omega)$, $\langle \mu, u \rangle \geq 0$. *Then* $\mu$ *is a positive measure.*

**Proof** We have to only verify that if $u_n \downarrow 0$, then $\langle \mu, u_n \rangle \to 0$. By the theorem of partition of unity, there exists $\psi \in \mathcal{D}(\Omega)$ such that $0 \leq \psi \leq 1$ and $\psi = 1$ on spt $u_0$. By the positivity of $\mu$, we obtain

$$0 \leq \langle \mu, ||u_n||_\infty \psi - u_n \rangle.$$

We conclude, using Dini's theorem, that

$$0 \leq \langle \mu, u_n \rangle \leq \langle \mu, \psi \rangle ||u_n||_\infty \to 0.$$                                     □

Let $\mu : \mathcal{K}(\Omega) \to \mathbb{R}$ be the difference of two positive measures $\mu_+$ and $\mu_-$. Then for every $u \in \mathcal{K}_+(\Omega)$,

$$\sup\{\langle \mu, f \rangle : f \in \mathcal{K}(\Omega), |f| \leq u\} \leq \langle \mu_+, u \rangle + \langle \mu_-, u \rangle < +\infty.$$

We shall prove the converse.

**Definition 5.1.6** Let $M \geq 1$. A measure is a linear functional $\mu : \mathcal{K}(\Omega; \mathbb{R}^M) \to \mathbb{R}$ such that for every $u \in \mathcal{K}_+(\Omega)$,

$$\langle |\mu|, u \rangle = \sup\{\langle \mu, f \rangle : f \in \mathcal{K}(\Omega; \mathbb{R}^M), |f| \leq u\} < +\infty.$$

The measure is scalar when $M = 1$ and vectorial when $M \geq 2$.

**Theorem 5.1.7** *Let $\mu : \mathcal{K}(\Omega; \mathbb{R}^M) \to \mathbb{R}$ be a measure. Then the functional defined on $\mathcal{K}(\Omega)$ by*

$$\langle |\mu|, u \rangle = \langle |\mu|, u^+ \rangle - \langle |\mu|, u^- \rangle$$

*is a positive measure.*

**Proof**

1. Let $u, v \in \mathcal{K}_+(\Omega)$, $f, g \in \mathcal{K}(\Omega; \mathbb{R}^M)$ be such that $|f| \leq u$ and $|g| \leq v$. Then

$$\langle \mu, f \rangle + \langle \mu, g \rangle = \langle \mu, f + g \rangle \leq \langle |\mu|, u + v \rangle.$$

Taking the supremum, we obtain

$$\langle |\mu|, u \rangle + \langle |\mu|, v \rangle \leq \langle |\mu|, u + v \rangle.$$

2. Let $u, v \in \mathcal{K}_+(\Omega)$, $h \in \mathcal{K}(\Omega; \mathbb{R}^M)$ be such that $|h| \leq u + v$. Define $f$ and $g$ on $\Omega$ by

$$f = uh/(u + v), g = vh/(u + v), \quad u + v > 0,$$
$$f = g = 0, \quad u + v = 0.$$

It is easy to verify that $f, g \in \mathcal{K}(\Omega; \mathbb{R}^M)$ and $|f| \leq u$, $|g| \leq v$, so that

$$\langle \mu, h \rangle = \langle \mu, f \rangle + \langle \mu, g \rangle \leq \langle |\mu|, u \rangle + \langle |\mu|, v \rangle.$$

Taking the supremum, we obtain

$$\langle |\mu|, u + v \rangle \leq \langle |\mu|, u \rangle + \langle |\mu|, v \rangle.$$

Hence, by the preceding step,

$$\langle |\mu|, u + v \rangle = \langle |\mu|, u \rangle + \langle |\mu|, v \rangle.$$

3. Let $u_k, v_k \in \mathcal{K}_+(\Omega)$, $k = 1, 2$, be such that $u_1 - v_1 = u_2 - v_2$. Then

$$\langle |\mu|, u_1 \rangle + \langle |\mu|, v_2 \rangle = \langle |\mu|, u_1 + v_2 \rangle = \langle |\mu|, u_2 + v_1 \rangle = \langle |\mu|, u_2 \rangle + \langle |\mu|, v_1 \rangle,$$

so that

$$\langle |\mu|, u_1 \rangle - \langle |\mu|, v_1 \rangle = \langle |\mu|, u_2 \rangle - \langle |\mu|, v_2 \rangle.$$

Since for every $u, v \in \mathcal{K}(\Omega)$,

$$(u + v)^+ - (u + v)^- = u + v = u^+ + v^+ - (u^- + v^-),$$

we conclude that

$$\langle |\mu|, u + v \rangle = \langle |\mu|, u \rangle + \langle |\mu|, v \rangle.$$

4. It is clear that for every $u \in \mathcal{K}(\Omega)$ and every $\lambda \in \mathbb{R}$,

$$\langle |\mu|, \lambda u \rangle = \lambda \langle |\mu|, u \rangle,$$

and that for $u \in \mathcal{K}_+(\Omega)$, $\langle |\mu|, u \rangle \geq 0$.                                    □

**Corollary 5.1.8 (Jordan Decomposition Theorem)** *Let* $\mu : \mathcal{K}(\Omega) \to \mathbb{R}$ *be a scalar measure. Then* $\mu = \mu_+ - \mu_-$, *where*

$$\mu_+ = \frac{|\mu| + \mu}{2}, \quad \mu_- = \frac{|\mu| - \mu}{2}$$

*are positive measures.*

We need a new function space.

**Definition 5.1.9** We define

$$C_0(\Omega) = \{u \in \mathcal{BC}(\Omega) : \text{for every } \varepsilon > 0, \text{ there exists a compact subset } K \text{ of } \Omega$$
$$\text{such that } \sup_{\Omega \setminus K} |u| < \varepsilon \Big\}.$$

*Example* The space $C_0(\mathbb{R}^N)$ is the set of continuous functions on $\mathbb{R}^N$ tending to 0 at infinity.

**Proposition 5.1.10** *The space* $C_0(\Omega)$ *is the closure of* $\mathcal{K}(\Omega)$ *in* $\mathcal{BC}(\Omega)$. *In particular,* $C_0(\Omega)$ *is separable.*

**Proof** If $u$ belongs to the closure of $\mathcal{K}(\Omega)$ in $\mathcal{BC}(\Omega)$, then for every $\varepsilon > 0$, there exists $v \in \mathcal{K}(\Omega)$ such that $||u - v||_\infty < \varepsilon$. Let $K = \text{spt } u$. We obtain

$$\sup_{\Omega \setminus K} |u(x)| = \sup_{\Omega \setminus K} |u(x) - v(x)| < \varepsilon.$$

If $u \in C_0(\Omega)$, then for every $\varepsilon > 0$, there exists a compact subset $K$ of $\Omega$ such that $\sup_{\Omega \setminus K} |u(x)| < \varepsilon$. The theorem of partitions of unity implies the existence of

$\varphi \in \mathcal{D}(\Omega)$ such that $0 \leq \varphi \leq 1$ and $\varphi = 1$ on $K$. Define $v = \varphi u$. Then $v \in \mathcal{K}(\Omega)$ and

$$||u - v||_\infty = \sup_{\Omega \setminus K} (1 - \varphi(x))|u(x)| < \varepsilon.$$

Hence, $C_0(\Omega)$ is the closure of $\mathcal{K}(\Omega)$ in $\mathcal{BC}(\Omega)$. By Propositions 2.3.2 and 3.3.11, $C_0(\Omega)$ is separable. $\qquad\square$

**Definition 5.1.11** The total variation of the measure $\mu$ : $\mathcal{K}(\Omega; \mathbb{R}^M) \to \mathbb{R}$ is defined by

$$||\mu||_\Omega = \sup\{\langle \mu, f \rangle : f \in \mathcal{K}(\Omega; \mathbb{R}^M), ||f||_\infty \leq 1\}.$$

The measure $\mu$ is finite if $||\mu||_\Omega < \infty$. By the preceding proposition, every finite measure $\mu$ has a continuous extension to $C_0(\Omega; \mathbb{R}^M)$. A sequence $(\mu_n)$ of finite measures converges weakly to $\mu$ if for every $f \in C_0(\Omega; \mathbb{R}^M)$,

$$\langle \mu_n, f \rangle \to \langle \mu, f \rangle.$$

**Theorem 5.1.12 (de la Vallée Poussin, 1932)** *Every sequence $(\mu_n)$ of measures on $\Omega$ such that* $\sup_n ||\mu_n||_\Omega < \infty$ *contains a weakly convergent subsequence.*

**Proof** By the preceding proposition, $C_0(\Omega; \mathbb{R}^M)$ is separable. It suffices then to use Banach's theorem. $\qquad\square$

## 5.2 James Representation Theorem

Let us define two useful classes of normed spaces.

**Definition 5.2.1** A normed space is smooth if its norm $F(u) = ||u||$ has a linear directional derivative $F'(u)$ for every $u \neq 0$:

$$\langle F'(u), v \rangle = \frac{d}{d\varepsilon}\Big|_{\varepsilon=0} F(u + \varepsilon v).$$

**Definition 5.2.2** A normed space is uniformly convex if for every $0 < \varepsilon \leq 1$,

$$\delta_X(\varepsilon) = \inf\left\{1 - ||\frac{u+v}{2}|| : ||u|| = ||v|| = 1, ||u - v|| \geq 2\varepsilon\right\} > 0.$$

The function $\delta_X(\varepsilon)$ is the modulus of convexity of the space.

The proof of the next result is left to the reader.

**Proposition 5.2.3** *Let $X$ be a smooth normed space and $u \in X \setminus \{0\}$. Then* $||F'(u)|| = 1$ *and*

$$\langle F'(u), u \rangle = ||u|| = \max_{\substack{f \in X^* \\ ||f||=1}} \langle f, u \rangle.$$

Choose $f \neq 0$ in the dual of the normed space $X$ and consider the dual problem

$$\begin{cases} \text{maximize } \langle f, u \rangle, \\ u \in X, ||u|| = 1. \end{cases} \tag{$\mathcal{P}$}$$

**Lemma 5.2.4** *Let $X$ be a smooth normed space, $f \in X^* \setminus \{0\}$, and $u$ a solution of* $(\mathcal{P})$. *Then* $f = ||f|| F'(u)$.

**Proof** By assumption, $\langle f, u \rangle = ||f||$. Let $v \in X$. The function

$$g(\varepsilon) = ||f|| \, ||u + \varepsilon v|| - \langle f, u + \varepsilon v \rangle$$

reaches its minimum at $\varepsilon = 0$. Hence, $g'(0) = 0$ and

$$||f|| \langle F'(u), v \rangle - \langle f, v \rangle = 0.$$

Since $v \in X$ is arbitrary, the proof is complete.                                        □

**Lemma 5.2.5** *Let $X$ be a uniformly convex Banach space and $f \in X^* \setminus \{0\}$. Then* $(\mathcal{P})$ *has a unique solution.*

**Proof** Let $(u_n) \subset X$ be a maximizing sequence for the problem $(\mathcal{P})$:

$$||u_n|| = 1, \quad \langle f, u_n \rangle \to ||f||, \quad n \to \infty.$$

Let us prove that $(u_n)$ is a Cauchy sequence. Let $0 < \varepsilon < 1$, and let $\delta_X(\varepsilon)$ be the modulus of convexity of $X$ at $\varepsilon$. There exists $m$ such that for $j, k \geq m$,

$$||f||(1 - \delta_X(\varepsilon)) < (\langle f, u_j \rangle + \langle f, u_k \rangle)/2 = \langle f, \frac{u_j + u_k}{2} \rangle \leq ||f|| \frac{||u_j + u_k||}{2}.$$

Hence, $j, k \geq m \implies ||u_j - u_k|| < 2\varepsilon$. Since $X$ is complete, $(u_n)$ converges to $u \in X$. By continuity, $||u|| = 1$ and $\langle f, u \rangle = ||f||$. Hence, $u$ is a solution of $(\mathcal{P})$.

Assume that $u$ and $v$ are solutions of $(\mathcal{P})$. The sequence $(u, v, u, v, \ldots)$ is maximizing. Hence, it is a Cauchy sequence, so that $u = v$.                          □

From the two preceding lemmas, we infer the *James representation theorem*.

**Theorem 5.2.6** *Let X be a smooth uniformly convex Banach space and $f \in X^* \setminus \{0\}$. Then there exists one and only one $u \in X$ such that*

$$\|u\| = 1, \quad \langle f, u \rangle = \|f\|, \quad f = \|f\| F'(u).$$

From the James representation theorem, we deduce a variant of the *Hahn–Banach theorem.*

**Theorem 5.2.7** *Let Y be a subspace of a smooth uniformly convex Banach space X and $f \in Y^*$. Then there exists one and only one $g \in X^*$ such that $\|g\| = \|f\|$ and $g\big|_Y = f$.*

**Proof** *Existence* If $f = 0$, then $g = 0$. Let $f \neq 0$. After extending $f$ to $\overline{Y}$ by Proposition 3.2.3, we can assume that $Y$ is closed.

The James representation theorem implies the existence of one and only one $u \in Y$ such that

$$\|u\| = 1, \quad \langle f, u \rangle = \|f\|, \quad f = \|f\| \left(F\big|_Y\right)'(u).$$

Define $g = \|f\| \, F'(u)$. It is clear that $\|g\| = \|f\|$ and

$$g\big|_Y = \|f\| \left(F\big|_Y\right)'(u) = f.$$

*Uniqueness* If $h \in X^*$ is such that $\|h\| = \|f\|$ and $h\big|_Y = f$, then

$$\langle h, u \rangle = \langle f, u \rangle = \|f\| = \|h\|.$$

Lemma 5.2.4 implies that $h = \|h\| F'(u) = \|f\| F'(u)$. □

## 5.3 Duality of Hilbert Spaces

By the Cauchy–Schwarz inequality, for every $g$ fixed in the Hilbert space $X$, the linear functional

$$X \to \mathbb{R} : v \mapsto (g|v)$$

is continuous. The *Fréchet–Riesz theorem* asserts that every continuous linear functional on $X$ has this representation.

**Theorem 5.3.1** *Let $X$ be a Hilbert space and $f \in X^*$. Then there exists one and only one $g \in X$ such that for every $v \in X$,*

$$\langle f, v \rangle = (g|v).$$

*Moreover, $||g|| = ||f||$.*

**Proof** *Existence* If $f = 0$, then $g = 0$. Assume that $f \neq 0$. It follows from the parallelogram identity that for $0 < \varepsilon \leq 1$, $\delta(\varepsilon) \geq 1 - \sqrt{1 - \varepsilon^2} > 0$. Hence, $X$ is uniformly convex.

If $u \in X \setminus \{0\}$, we find that

$$\langle F'(u), v \rangle = \frac{d}{d\varepsilon}\Big|_{\varepsilon=0} ||u + \varepsilon v|| = ||u||^{-1}(u|v).$$

Hence, $X$ is smooth.

The James representation theorem implies the existence of $u \in X$ such that

$$||u|| = 1, \quad \langle f, u \rangle = ||f||, \quad f = ||f||F'(u).$$

But then, for every $v \in X$,

$$\langle f, v \rangle = ||f||(u|v) = (||f||u|v).$$

*Uniqueness*  If for every $v \in X$,

$$(g|v) = \langle f, v \rangle = (h|v),$$

then $||g - h||^2 = 0$ and $g = h$.                                                    □

**Definition 5.3.2** The vector space $X$ is the direct sum of the subspaces $Y$ and $Z$ if $Y \cap Z = \{0\}$ and $X = \{y + z : y \in Y, z \in Z\}$. We then write $X = Y \oplus Z$, and every $u \in X$ has a unique decomposition $u = y + z$, $y \in Y$, $z \in Z$.

**Definition 5.3.3** The orthogonal space to a subset $Y$ of a pre-Hilbert space $X$ is defined by

$$Y^\perp = \{z \in X : \text{for every } y \in Y, (z|y) = 0\}.$$

It is easy to verify that $Y^\perp$ is a closed subspace of $X$.

**Corollary 5.3.4** *Let $Y$ be a closed subspace of a Hilbert space $X$. Then $X$ is the direct sum of $Y$ and $Y^\perp$.*

**Proof** If $u \in Y \cap Y^{\perp}$, then $(u|u) = 0$ and $u = 0$.

Let $u \in X$. The Fréchet–Riesz theorem implies the existence of $y \in Y$ such that for every $v \in Y$, $(u|v) = (y|v)$. But then $z = u - y \in Y^{\perp}$.  □

**Corollary 5.3.5**  *A subspace $Y$ of a Hilbert space $X$ is dense if and only if $Y^{\perp} = \{0\}$.*

**Proof** Let $Y$ be a subspace of $X$. Then $\overline{Y}$ is a closed subspace of $X$. By continuity of the scalar product, $Y^{\perp} = \overline{Y}^{\perp}$. It follows from the preceding corollary that

$$X = \overline{Y} \oplus \overline{Y}^{\perp} = \overline{Y} \oplus Y^{\perp}.$$

□

**Definition 5.3.6**  A sequence $(u_n)$ converges weakly to $u$ in the Hilbert space $X$ if for every $v \in X$, $(u_n|v) \to (u|v)$. We then write $u_n \rightharpoonup u$.

**Proposition 5.3.7**  *Let $Z$ be a dense subset of a Hilbert space $X$ and $(u_n) \subset X$ be such that:*

*(a)* $\sup_n \|u_n\| < \infty$;

*(b)* *for every $v \in Z$, $(u_n|v)$ converges.*

*Then $(u_n)$ converges weakly in $X$.*

**Proof** It suffices to use Proposition 5.1.2 and the Fréchet–Riesz theorem.  □

**Theorem 5.3.8**  *Let $(u_n)$ be a sequence weakly convergent to $u$ in the Hilbert space $X$. Then $(u_n)$ is bounded and*

$$\|u\| \le \varliminf_{n \to \infty} \|u_n\|.$$

**Proof** It suffices to use Theorem 5.1.3 and the Fréchet–Riesz theorem.  □

**Theorem 5.3.9**  *Every bounded sequence in a Hilbert space contains a weakly convergent subsequence.*

**Proof** Let $(u_n)$ be a bounded sequence in the Hilbert space $X$, and let $Y$ be the closure of the space generated by $(u_n)$. The sequence $(u_n)$ is bounded in the separable Hilbert space $Y$. By the Banach theorem and the Fréchet–Riesz theorem, there exists a subsequence $v_k = u_{n_k}$ weakly converging to $u$ in $Y$. For every $v \in X$, $v = y + z$, $y \in Y$, and $z \in Y^{\perp}$ by Corollary 5.3.4. By definition, $(v_k|z) = (u|z) = 0$. Hence, $(v_k|v) \to (u|v)$ and $v_k \rightharpoonup u$ in $X$.  □

**Definition 5.3.10**  Let $\mu : \mathcal{L} \to \mathbb{R}$ and $\nu : \mathcal{L} \to \mathbb{R}$ by positive measures on $\Omega$. By definition, $\mu \le \nu$ if for every $u \in \mathcal{L}$, $u \ge 0$, $\displaystyle\int_{\Omega} u \, d\mu \le \int_{\Omega} u \, d\nu$.

**Lemma 5.3.11** *Let* $\mu \leq \nu$. *Then* $L^1(\Omega, \nu) \subset L^1(\Omega, \mu)$, *and for every* $u \in$ $L^1(\Omega, \nu)$, $||u||_{L^1(\Omega, \mu)} \leq ||u||_{L^1(\Omega, \nu)}$.

***Proof*** Let $u \in L^1(\Omega, \nu)$. By the density theorem, there exists a sequence $(u_n) \subset \mathcal{L}$ such that $u_n \to u$ in $L^1(\Omega, \nu)$ and $\nu$-almost everywhere. Clearly, $u_n \to u$ $\mu$-almost everywhere. By Fatou's lemma, $u \in L^1(\Omega, \mu)$ and

$$\int_\Omega |u| d\mu \leq \lim_{n\to\infty} \int_\Omega |u_n| d\mu \leq \lim_{n\to\infty} \int_\Omega |u_n| d\nu = \int_\Omega |u| d\nu. \qquad \square$$

**Lemma 5.3.12 (von Neumann)** *Let* $\mu \leq \nu$ *and* $\nu(\Omega) < +\infty$. *Then there exists one and only one function* $g : \Omega \to [0, 1]$ *measurable with respect to* $\nu$ *and such that for every* $u \in L^1(\Omega, \nu)$,

$$\int_\Omega u \, d\mu = \int_\Omega ug \, d\nu.$$

***Proof*** By assumption, $L^2(\Omega, \nu) \subset L^1(\Omega, \nu)$. Let us define $f$ on $L^2(\Omega, \nu)$ by

$$\langle f, u \rangle = \int_\Omega u \, d\mu.$$

By the Cauchy–Schwarz inequality, we have

$$|\langle f, u \rangle| \leq (\mu(\Omega))^{1/2} \left( \int_\Omega u^2 d\mu \right)^{1/2} \leq (\mu(\Omega))^{1/2} \left( \int_\Omega u^2 d\nu \right)^{1/2}.$$

The Fréchet–Riesz theorem implies the existence of one and only one function $g \in L^2(\Omega, \nu)$ such that for every $v \in L^2(\Omega, \nu)$,

$$\int_\Omega v \, d\mu = \int_\Omega vg \, d\nu. \qquad (*)$$

In particular, we obtain

$$0 \leq \int_\Omega g^- d\mu = -\int_\Omega (g^-)^2 d\nu$$

and $\nu(\{g < 0\}) = 0$. Similarly, we have

$$0 \leq \int_\Omega (1 - g)^- d\nu - \int_\Omega (1 - g)^- d\mu = -\int_\Omega \left[ (1 - g)^- \right]^2 d\nu$$

and $\nu(\{g > 1\}) = 0$. Let $u \in L^1(\Omega, \nu)$, $u \geq 0$, and define $u_n = \min(u, n)$. We deduce from $(*)$ and Levi's theorem that

$$\int_\Omega u\, d\mu = \lim_{n\to\infty} \int_\Omega u_n d\mu = \lim_{n\to\infty} \int_\Omega u_n g\, dv = \int_\Omega ug\, dv.$$

Since $u = u^+ - u^-$, the preceding equality holds for every $u \in L^1(\Omega, v)$.   □

Let us prove *Lebesgue's decomposition theorem*.

**Theorem 5.3.13** *Let $\mu : \mathcal{L} \to \mathbb{R}$ and $v : \mathcal{L} \to \mathbb{R}$ be positive measures on $\Omega$ such that $\mu(\Omega) < \infty$, $v(\Omega) < \infty$. Then there exist $h \in L^1(\Omega, v)$ and $\Sigma \subset \Omega$, measurable with respect to $\mu$ and $v$, such that:*

*(a)* $v(\Sigma) = 0$, $h \geq 0$;
*(b)* *for every* $u \in L^1(\Omega, \mu) \cap L^1(\Omega, v)$, $uh \in L^1(\Omega, v)$ *and*

$$\int_\Omega u\, d\mu = \int_\Omega uh\, dv + \int_\Sigma u\, d\mu.$$

*Proof* Let $X = L^1(\Omega, \mu) \cap L^1(\Omega, v)$. The preceding lemma implies the existence of $g : \Omega \to [0, 1]$, measurable with respect to $\mu$ and $v$, such that for every $v \in X$,

$$\int_\Omega v\, d\mu = \int_\Omega vg\, d\mu + \int_\Omega vg\, dv.$$

Let $\Sigma = \{g = 1\}$. Since for every $v \in X$,

$$\int_\Omega v(1 - g)d\mu = \int_\Omega vg\, dv, \tag{$*$}$$

we obtain $v(\Sigma) = 0$. Let us define $h = \chi_{\Omega \setminus \Sigma} g/(1 - g)$. Choose $u \in X$, $u \geq 0$, and define

$$u_n = (1 + g + \ldots + g^n)u.$$

We deduce from $(*)$ and Levi's theorem that

$$\int_{\Omega \setminus \Sigma} u\, d\mu = \int_{\Omega \setminus \Sigma} ug/(1 - g)dv = \int_\Omega uh\, dv.$$

Since $u = u^+ - u^-$, the preceding equality holds for every $u \in X$. Finally, we have

$$\int_\Omega h\, dv = \mu(\Omega \setminus \Sigma) < +\infty.$$   □

*Remark* Every other decomposition of $\mu$ corresponding to $h_0$ and $\Sigma_0$ is such that $\mu(\Sigma_0 \setminus \Sigma) = \mu(\Sigma \setminus \Sigma_0) = 0$ and $v(\{h_0 \neq h\}) = 0$.

Let us prove the *polar decomposition of vector measures theorem.*

**Theorem 5.3.14** *Let $\Omega$ be an open subset of $\mathbb{R}^N$, and let $\mu : \mathcal{K}(\Omega; \mathbb{R}^M) \to \mathbb{R}$ be a measure such that $||\mu||_\Omega < +\infty$. Then there exists a function $g : \Omega \to \mathbb{R}^M$ such that:*

*(a)  $g$ is $|\mu|$-measurable;*
*(b)  $|g(x)| = 1$, $|\mu|$-almost everywhere on $\Omega$;*
*(c)  for all $f \in \mathcal{K}(\Omega; \mathbb{R}^M)$, $\langle \mu, f \rangle = \int_\Omega f \cdot g \, d|\mu|$.*

**Proof** Let $e_1, \ldots, e_M$ be the standard basis of $\mathbb{R}^M$, and for $1 \leq k \leq M$, define $\mu_k$ on $\mathcal{K}(\Omega)$ by

$$\langle \mu_k, u \rangle = \langle \mu, u \, e_k \rangle.$$

It is clear that for all $u \in \mathcal{K}(\Omega)$,

$$|\langle \mu_k, u \rangle| \leq \int_\Omega |u| \, d|\mu| \leq ||\mu||_\Omega^{1/2} \, ||u||_{L^2(\Omega, |\mu|)}.$$

Since $\mathcal{K}(\Omega)$ is dense in $L^2(\Omega, |\mu|)$, Proposition 3.2.3 implies the existence of a continuous extension of $\mu_k$ to $L^2(\Omega, |\mu|)$. By the Fréchet–Riesz representation theorem, there exists $g_k \in L^2(\Omega, |\mu|)$ such that for all $u \in \mathcal{K}(\Omega)$,

$$\langle \mu_k, u \rangle = \int_\Omega u \, g_k \, d|\mu|.$$

We define $g = \sum_{k=1}^M g_k e_k$, so that for all $f \in \mathcal{K}(\Omega; \mathbb{R}^M)$,

$$\langle \mu, f \rangle = \sum_{k=1}^M \langle \mu, f_k e_k \rangle = \sum_{k=1}^M \int_\Omega f_k g_k \, d|\mu| = \int_\Omega f \cdot g \, d|\mu|.$$

Let $u \in \mathcal{K}_+(\Omega)$. We have, by Definition 5.1.6,

$$c = \sup \left\{ \int_\Omega f \cdot g \, d|\mu| : f \in \mathcal{K}(\Omega; \mathbb{R}^M), |f| \leq u \right\} = \int_\Omega u \, d|\mu|.$$

It is clear that $c \leq \int u|g| \, d|\mu|$. Theorem 4.2.11 implies the existence of $(w_n) \subset \mathcal{K}(\Omega; \mathbb{R}^M)$ converging to $g$ in $L^2(\Omega, |\mu|)$. Let us define

$$v_n = u \, w_n / \sqrt{|w_n|^2 + 1/n}.$$

We infer from Lebesgue's dominated convergence theorem that

$$c \leq \int_{\Omega} u|g| \, d|\mu| = \lim_{n \to \infty} \int_{\Omega} v_n \cdot g \, d|\mu| \leq c.$$

We conclude that for all $u \in \mathcal{K}(\Omega)$,

$$\int_{\Omega} u|g| \, d|\mu| = \int_{\Omega} u \, d|\mu|.$$

Hence, $|g| - 1$ is orthogonal to $\mathcal{K}(\Omega)$ in $L^2(\Omega, |\mu|)$. By Corollary 5.3.5, $|g| - 1 = 0$, $|\mu|$-almost everywhere. □

## 5.4  Duality of Lebesgue Spaces

Let $1 < p < \infty$, and let $p'$ be the exponent conjugate to $p$ defined by $1/p + 1/p' = 1$. By Hölder's inequality, for every $g$ fixed in $L^{p'}(\Omega, \mu)$, the linear functional

$$L^p(\Omega, \mu) \to \mathbb{R} : v \mapsto \int_{\Omega} gv \, d\mu$$

is continuous. Riesz's representation theorem asserts that every continuous linear functional on $L^p(\Omega, \mu)$ has this representation. We denote by $\mu : \mathcal{L} \to \mathbb{R}$ a positive measure on $\Omega$.

**Theorem 5.4.1** *Let* $1 < p < \infty$. *Then the space* $L^p(\Omega, \mu)$ *is smooth, and the directional derivative of the norm* $F(u) = ||u||_p$ *is given, for* $u \neq 0$, *by*

$$\langle F'(u), v \rangle = ||u||_p^{1-p} \int_{\Omega} |u|^{p-2} uv \, d\mu.$$

**Proof** We define $G(u) = \int_{\Omega} |u|^p d\mu$, and we choose $u, v \in L^p$. By the fundamental theorem of calculus, for $0 < |\varepsilon| < 1$ and almost all $x \in \Omega$,

$$\left| |u(x) + \varepsilon v(x)|^p - |u(x)|^p \right| \leq p \left| \int_0^{\varepsilon} |u(x) + tv(x)|^{p-1} |v(x)| dt \right|$$

$$\leq p|\varepsilon| \Big( |u(x)| + |v(x)| \Big)^{p-1} |v(x)|.$$

It follows from Hölder's inequality that

$$(|u(x)| + |v(x)|)^{p-1} |v(x)| \in L^1.$$

Lebesgue's dominated convergence theorem ensures that

$$\langle G'(u), v \rangle = \frac{d}{d\varepsilon}\Big|_{\varepsilon=0} G(u + \varepsilon v) = p \int_{\Omega} |u|^{p-2} uv \, d\mu.$$

Hence for $u \neq 0$,

$$\langle F'(u), v \rangle = \frac{d}{d\varepsilon}\Big|_{\varepsilon=0} \sqrt[p]{G(u + \varepsilon v)} = G(u)^{\frac{1-p}{p}} \int_{\Omega} |u|^{p-2} uv \, d\mu. \qquad \square$$

**Theorem 5.4.2 (Clarkson, 1936)** *Let* $1 < p < \infty$. *Then the space* $L^p(\Omega, \mu)$ *is uniformly convex.*

**Proof** If $L^p$ is not uniformly convex, then there exist $0 < \varepsilon \leq 1$ and $(u_n)$, $(v_n)$ such that

$$||u_n||_p = ||v_n||_p = 1, \quad ||u_n - v_n||_p \to 2\varepsilon \quad \text{and} \quad ||u_n + v_n||_p \to 2.$$

If $2 \leq p < \infty$, we deduce from Hanner's inequality that

$$||u_n + v_n||_p^p + ||u_n - v_n||_p^p \leq 2^p.$$

Taking the limit, we obtain $2^p + 2^p \varepsilon^p \leq 2^p$. This is a contradiction.

If $1 < p \leq 2$, we deduce from Hanner's inequality that

$$\left( \left\| \frac{u_n + v_n}{2} \right\|_p + \left\| \frac{u_n - v_n}{2} \right\|_p \right)^p + \left| \left\| \frac{u_n + v_n}{2} \right\|_p - \left\| \frac{u_n - v_n}{2} \right\|_p \right|^p \leq 2.$$

Taking the limit, we find by strict convexity that

$$2 < (1 + \varepsilon)^p + (1 - \varepsilon)^p \leq 2.$$

This is also a contradiction. $\qquad \square$

**Theorem 5.4.3 (Riesz's Representation Theorem)** *Let* $1 < p < \infty$ *and* $f \in (L^p(\Omega, \mu))^*$. *Then there exists one and only one* $g \in L^{p'}(\Omega, \mu)$ *such that for every* $v \in L^p(\Omega, \mu)$,

$$\langle f, v \rangle = \int_{\Omega} gv \, d\mu.$$

*Moreover,* $||g||_{p'} = ||f||$.

**Proof** Existence. If $f = 0$, then $g = 0$. Assume $f \neq 0$. Since $L^p$ is smooth and uniformly convex by the preceding theorems, the James representation theorem implies the existence of $u \in L^p$ such that

$$||u||_p = 1, \quad \langle f, u \rangle = ||f||, \quad f = ||f||F'(u).$$

But then for every $v \in L^p$,

$$\langle f, v \rangle = ||f|| \int_\Omega |u|^{p-2} uv \, d\mu.$$

Define $g = ||f|| \, |u|^{p-2} u$. It is easy to verify that $g \in L^{p'}$ and $||g||_{p'} = ||f||$.

*Uniqueness* It suffices to prove that if $g \in L^{p'}$ is such that for every $v \in L^p$, $\int_\Omega gv \, d\mu = 0$, then $g = 0$. Since $|g|^{p'-2} g \in L^p$, we obtain

$$||g||_{p'}^{p'} = \int_\Omega |g|^{p'} d\mu = 0. \qquad \square$$

**Definition 5.4.4** Let $1 < p < \infty$. We identify the spaces $(L^{p'}(\Omega, \mu))^*$ and $L^p(\Omega, \mu)$. A sequence $(u_n)$ converges weakly to $u$ in $L^p(\Omega, \mu)$ if for every $v \in L^{p'}(\Omega, \mu)$,

$$\int_\Omega u_n v \, d\mu \rightarrow \int_\Omega uv \, d\mu.$$

We then write $u_n \rightharpoonup u$.

**Proposition 5.4.5** *Let $1 < p < \infty$, let $Z$ be a dense subset of $L^{p'}(\Omega, \mu)$, and let $(u_n) \subset L^p(\Omega, \mu)$ be such that:*

*(a)* $\sup_n ||u_n||_p < \infty$;

*(b) for every $v \in Z$, $\int_\Omega u_n v \, d\mu$ converges.*

*Then $(u_n)$ converges weakly to $u \in L^p(\Omega, \mu)$.*

**Proof** It suffices to use Proposition 5.1.2. $\qquad \square$

**Theorem 5.4.6** *Let $1 < p < \infty$, and let $(u_n)$ be a sequence weakly convergent to $u$ in $L^p(\Omega, \mu)$. Then $(u_n)$ is bounded and*

$$||u||_p \leq \varliminf_{n \to \infty} ||u_n||_p.$$

**Proof** It suffices to use Theorem 5.1.3. $\qquad \square$

**Proposition 5.4.7** *Let* $1 < p < \infty$, *and let* $(u_n) \subset L^p(\Omega, \mu)$ *be such that:*

*(a)* $c = \sup \|u_n\|_p < \infty$;
*(b)* $(u_n)$ *converges almost everywhere to* $u$ *on* $\Omega$.

*Then* $u_n \rightharpoonup u$ *in* $L^p(\Omega, \mu)$.

**Proof** By Fatou's lemma, $\|u\|_p \le c$. We choose $v$ in $L^{p'}(\Omega, \mu)$, and we define

$$A_n = \{x \in \Omega : |u_n(x) - u(x)| \le |v(x)|^{p'-1}\}, \quad B_n = \Omega \setminus A_n.$$

We deduce from Hölder's and Minkowski's inequalities that

$$\int_\Omega |u_n - u|\, |v| d\mu \le \int_{A_n} |u_n - u|\, |v| d\mu + \|u_n - u\|_p \left(\int_{B_n} |v|^{p'} d\mu\right)^{1/p'}$$

$$\le \int_{A_n} |u_n - u|\, |v| d\mu + 2c \left(\int_{B_n} |v|^{p'} d\mu\right)^{1/p'}.$$

Lebesgue's dominated convergence theorem ensures that

$$\lim_{n \to \infty} \int_{A_n} |u_n - u|\, |v| d\mu = 0 = \lim_{n \to \infty} \int_{B_n} |v|^{p'} d\mu. \qquad \square$$

**Theorem 5.4.8** *Let* $1 < p < \infty$, *and let* $\Omega$ *be an open subset of* $\mathbb{R}^N$. *Then every bounded sequence in* $L^p(\Omega)$ *contains a weakly convergent subsequence.*

**Proof** By Theorem 4.2.11, $L^{p'}(\Omega)$ is separable. It suffices then to use Banach's theorem. $\qquad \square$

*Examples (Weak Convergence in* $L^p$*)* What are the obstructions to the (strong) convergence of weakly convergent sequences? We consider three processes by which in $L^p(\Omega)$,

$$u_n \rightharpoonup 0, \quad u_n \nrightarrow 0.$$

*Oscillation* The sequence $u_n(x) = \sqrt{\frac{2}{\pi}} \sin n\, x$ is orthonormal in $L^2(]0, \pi[)$. It follows from Bessel's inequality that $u_n \rightharpoonup 0$. But $\|u_n\|_2 = 1$.

*Concentration* Let $1 < p < \infty$, $u \in \mathcal{K}(\mathbb{R}^N) \setminus \{0\}$, and $u_n(x) = n^{N/p}u(nx)$. For every $n$, $\|u_n\|_p = \|u\|_p > 0$, and for all $x \neq 0$, $u_n(x) \to 0$, $n \to \infty$. By Proposition 5.4.6, $u_n \rightharpoonup 0$ in $L^p(\mathbb{R}^N)$.

*Translation*  Let $1 < p < \infty$, $u \in \mathcal{K}(\mathbb{R}^N) \setminus \{0\}$, and $u_n(x) = u(x_1 - n, x_2, \dots, x_N)$. For every $n$, $\|u_n\|_p = \|u\|_p > 0$, and for all $x$, $u_n(x) \to 0$, $n \to \infty$. By Proposition 5.4.7, $u_n \rightharpoonup 0$ in $L^p(\mathbb{R}^N)$.

## 5.5  Comments

A representation theorem gives to an abstract mathematical object like a functional a more concrete representation involving in many cases an integral. It replaces a structural definition by an analytic description. The first representation theorem (proved by Riesz in 1909 [61]) asserts that every continuous linear functional on $C([0, 1])$ is representable by a Stieltjes integral (see Sect. 10.1). In this chapter, we use as a basic tool the James representation theorem [35].

## 5.6  Exercises for Chap. 5

1. Define a sequence $(f_n)$ of finite measures on $]0, 1[$ such that:

   (a) $\|f_n\| \to \|f\|$;
   (b) $f_n \rightharpoonup f$;
   (c) $\|f_n - f\| \nrightarrow 0$.

2. Let $X$ be a Hilbert space, and let $(u_n) \subset X$ be such that:

   (a) $\overline{\lim} \|u_n\| \le \|u\|$;
   (b) $u_n \rightharpoonup u$.

   Then $\|u_n - u\| \to 0$.

3. Let $1 < p < \infty$ and $(u_n) \subset L^p(\Omega, \mu)$ be such that:

   (a) $\overline{\lim} \|u_n\|_p \le \|u\|_p$;
   (b) $u_n \rightharpoonup u$.

   Then $\|u_n - u\|_p \to 0$. *Hint*: If $v_n \rightharpoonup v$, then $\|v\|_p \le \underline{\lim} \|\frac{v_n + v}{2}\|_p$.

4. Let $1 < p < \infty$ and $u_n \rightharpoonup u$ in $L^p(\Omega, \mu)$. Is it true that

$$\lim_{n \to \infty} (\|u_n\|_p^p - \|u_n - u\|_p^p) = \|u\|_p^p \ ?$$

   *Hint*: When $p \ne 2$, construct a counterexample using oscillating step functions.

5. Let $X$ be a smooth uniformly convex Banach space and $f, g \in X^*$. Then

$$\max_{\substack{\langle g, y \rangle = 0 \\ \|y\| = 1}} \langle f, y \rangle = \min_{\lambda \in \mathbb{R}} \|f - \lambda g\|.$$

6. Let $C$ be a closed convex subset of a uniformly convex Banach space $X$. Then for every $u \in X$, there exists one and only one $v \in C$ such that $||u - v|| = d(u, C)$.

7. Let $\Omega$ be an open subset of $\mathbb{R}^N$ and $f \in L^1_{\text{loc}}(\Omega)$. Prove that

$$\mu : \mathcal{K}(\Omega) \to \mathbb{R} : u \mapsto \int_\Omega f(x)u(x)\, dx$$

is a measure on $\Omega$ such that

$$||\mu||_\Omega = \int_\Omega |f(x)|dx.$$

8. Let $\mu$ be a positive measure on $\Omega$ such that $\mu(\Omega) = 1$. We define, on $\mathcal{M}(\Omega, \mu)$,

$$||u||_\infty = \inf\{c \geq 0 : \text{almost everywhere on } \Omega, |u(x)| \leq c\}.$$

We also define

$$L^\infty(\Omega, \mu) = \{u \in \mathcal{M}(\Omega, \mu) : ||u||_\infty < +\infty\}.$$

We identify two functions of $L^\infty(\Omega, \mu)$ when they are $\mu$-almost everywhere equal. If $||u||_\infty < +\infty$, then $u \in \bigcap_{1 \leq p < \infty} L^p(\Omega, \mu)$ and $||u||_\infty = \lim_{p \to \infty} ||u||_p$.

9. Assume that $\mu(\Omega) = 1$. For every $f \in (L^1(\Omega, \mu))^*$, there exists one and only one $g \in L^\infty(\Omega, \mu)$ such that for every $v \in L^1(\Omega, \mu)$,

$$\langle f, v \rangle = \int_\Omega gv\, d\mu.$$

Moreover, $||g||_\infty = ||f||$. *Hint*: Use Riesz's representation theorem on $(L^p)^*$, $1 < p < \infty$.

10. Let $\Omega$ be a bounded open subset of $\mathbb{R}^N$, and let $(g_n) \subset L^\infty(\Omega, \Lambda_N)$ be such that $\sup_n ||g_n||_\infty < +\infty$. Then there exist a subsequence $(g_{n_k})$ of $(g_n)$ and $g \in L^\infty(\Omega, \Lambda_N)$ such that for every $v \in L^1(\Omega, \Lambda_N)$,

$$\lim_{k \to \infty} \int_\Omega g_{n_k} v\, dx = \int_\Omega gv\, dx.$$

# Chapter 6
# Sobolev Spaces

## 6.1 Weak Derivatives

Throughout this chapter, we denote by $\Omega$ an open subset of $\mathbb{R}^N$. We begin with an elementary computation.

**Lemma 6.1.1** *Let $1 \leq |\alpha| \leq m$ and let $f \in C^m(\Omega)$. Then for every $u \in C^m(\Omega) \cap \mathcal{K}(\Omega)$,*

$$\int_\Omega f \, D^\alpha u \, dx = (-1)^{|\alpha|} \int_\Omega (D^\alpha f)u \, dx.$$

**Proof** We assume that $\alpha = (0, \ldots, 0, 1)$. Let $u \in C^1(\Omega) \cap \mathcal{K}(\Omega)$, and define

$$g(x) = f(x)u(x), \quad x \in \Omega,$$
$$= 0, \quad\quad\quad x \in \mathbb{R}^N \setminus \Omega.$$

The fundamental theorem of calculus implies that for every $x' \in \mathbb{R}^{N-1}$,

$$\int_\mathbb{R} D^\alpha g(x', x_N) dx_N = 0.$$

Fubini's theorem ensures that

$$\int_\Omega (f D^\alpha u + (D^\alpha f)u)dx = \int_{\mathbb{R}^N} D^\alpha g \, dx = \int_{\mathbb{R}^{N-1}} dx' \int_\mathbb{R} D^\alpha g \, dx_N = 0.$$

When $|\alpha| = 1$, the proof is similar. It is easy to conclude the proof by induction. $\square$

Weak derivatives were defined by S.L. Sobolev in 1938.

© Springer Nature Switzerland AG 2022
M. Willem, *Functional Analysis*, Cornerstones,
https://doi.org/10.1007/978-3-031-09149-0_6

**Definition 6.1.2** Let $\alpha \in \mathbb{N}^N$ and $f \in L^1_{loc}(\Omega)$. By definition, the weak derivative of order $\alpha$ of $f$ exists if there is $g \in L^1_{loc}(\Omega)$ such that for every $u \in \mathcal{D}(\Omega)$,

$$\int_\Omega f \, D^\alpha u \, dx = (-1)^{|\alpha|} \int_\Omega gu \, dx.$$

The function $g$, if it exists, will be denoted by $\partial^\alpha f$.

By the annulation theorem, the weak derivatives are well defined.

**Proposition 6.1.3** *Assume that $\partial^\alpha f$ exists. On*

$$\Omega_n = \{x \in \Omega : d(x, \partial\Omega) > 1/n\},$$

*we have that*

$$D^\alpha(\rho_n * f) = \rho_n * \partial^\alpha f.$$

**Proof** We deduce from Proposition 4.3.6 and from the preceding definition that for every $x \in \Omega_n$,

$$D^\alpha(\rho_n * f)(x) = \int_\Omega D^\alpha_x \rho_n(x - y) f(y) dy$$

$$= (-1)^{|\alpha|} \int_\Omega D^\alpha_y \rho_n(x - y) f(y) dy$$

$$= (-1)^{2|\alpha|} \int_\Omega \rho_n(x - y) \partial^\alpha f(y) dy$$

$$= \rho_n * \partial^\alpha f(x). \qquad \square$$

**Theorem 6.1.4 (du Bois–Reymond Lemma)** *Let $|\alpha| = 1$ and let $f \in C(\Omega)$ be such that $\partial^\alpha f \in C(\Omega)$. Then $D^\alpha f$ exists and $D^\alpha f = \partial^\alpha f$.*

**Proof** By the preceding proposition, we have

$$D^\alpha(\rho_n * f) = \rho_n * \partial^\alpha f.$$

The fundamental theorem of calculus implies then that

$$\rho_n * f(x + \varepsilon\alpha) = \rho_n * f(x) + \int_0^\varepsilon \rho_n * \partial^\alpha f(x + t\alpha) dt.$$

By the regularization theorem,

$$\rho_n * f \to f, \qquad \rho_n * \partial^\alpha f \to \partial^\alpha f$$

uniformly on every compact subset of $\Omega$. Hence we obtain

$$f(x + \varepsilon\alpha) = f(x) + \int_0^\varepsilon \partial^\alpha f(x + t\alpha)dt,$$

so that $\partial^\alpha f = D^\alpha f$ by the fundamental theorem of calculus. □

*Notation* From now on, the derivatives of a continuously differentiable function will also be denoted by $\partial^\alpha$.

Let us prove the *closing lemma*. The *graph* of the weak derivative is closed in $L^1_{\text{loc}} \times L^1_{\text{loc}}$.

**Lemma 6.1.5** *Let* $(f_n) \subset L^1_{\text{loc}}(\Omega)$ *and let* $\alpha \in \mathbb{N}^N$ *be such that in* $L^1_{\text{loc}}(\Omega)$,

$$f_n \to f, \qquad \partial^\alpha f_n \to g.$$

*Then* $g = \partial^\alpha f$.
**Proof** For every $u \in \mathcal{D}(\Omega)$, we have by definition that

$$\int_\Omega f_n \partial^\alpha u \, dx = (-1)^{|\alpha|} \int_\Omega (\partial^\alpha f_n)u \, dx.$$

Since by assumption,

$$\left| \int_\Omega (f_n - f)\partial^\alpha u \, dx \right| \le ||\partial^\alpha u||_\infty \int_{\text{spt } u} |f_n - f|dx \to 0$$

and

$$\left| \int_\Omega (\partial^\alpha f_n - g)u \, dx \right| \le ||u||_\infty \int_{\text{spt } u} |\partial^\alpha f_n - g|dx \to 0,$$

we obtain

$$\int_\Omega f \partial^\alpha u \, dx = (-1)^{|\alpha|} \int_\Omega gu \, dx. \qquad \square$$

*Example (Weak Derivative)* If $-N < \lambda \le 1$, the function $f(x) = |x|^\lambda$ is locally integrable on $\mathbb{R}^N$. We approximate $f$ by

$$f_\varepsilon(x) = \left(|x|^2 + \varepsilon\right)^{\lambda/2}, \qquad \varepsilon > 0.$$

Then $f_\varepsilon \in C^\infty(\mathbb{R}^N)$ and

$$\partial_k f_\varepsilon(x) = \lambda \, x_k \left( |x|^2 + \varepsilon \right)^{\frac{\lambda-2}{2}},$$

$$\left| \partial_k f_\varepsilon(x) \right| \le \lambda |x|^{\lambda-1}.$$

If $\lambda > 1 - N$, we obtain in $L^1_{\mathrm{loc}}(\mathbb{R}^N)$ that

$$f_\varepsilon(x) \to f(x) = |x|^\lambda,$$

$$\partial_k f_\varepsilon(x) \to g(x) = \lambda \, x_k |x|^{\lambda-2}.$$

Hence $\partial_k f(x) = \lambda \, |x|^{\lambda-2} x_k$.

**Definition 6.1.6** The *gradient* of the (weakly) differentiable function $u$ is defined by

$$\nabla u = (\partial_1 u, \ldots, \partial_N u).$$

The *divergence* of the (weakly) differentiable vector field $v = (v_1, \ldots, v_N)$ is defined by

$$\mathrm{div}\, v = \partial_1 v_1 + \ldots + \partial_N v_N.$$

Let $1 \le p < \infty$ and $u \in L^1_{\mathrm{loc}}(\Omega)$ be such that $\partial_j u \in L^p(\Omega)$, $j = 1, \ldots, N$. We define

$$||\nabla u||_{L^p(\Omega)} = \left( \int_\Omega |\nabla u|^p dx \right)^{1/p} = \left( \int_\Omega \left| \sum_{j=1}^N (\partial_j u)^2 \right|^{p/2} dx \right)^{1/p}.$$

**Theorem 6.1.7** *Let $1 < p < \infty$ and let $(u_n) \subset L^1_{\mathrm{loc}}(\Omega)$ be such that*

*(a)  $u_n \to u$ in $L^1_{\mathrm{loc}}(\Omega)$;*
*(b)  for every $n$, $\nabla u_n \in L^p(\Omega; \mathbb{R}^N)$;*
*(c)  $c = \sup_n ||\nabla u_n||_p < \infty$.*

*Then $\nabla u \in L^p(\Omega; \mathbb{R}^N)$ and*

$$||\nabla u||_p \le \varliminf_{n \to \infty} ||\nabla u_n||_p.$$

**Proof** We define $f$ on $\mathcal{D}(\Omega; \mathbb{R}^N)$ by

$$\langle f, v \rangle = \int_\Omega u \, \mathrm{div} \, v \, dx.$$

We have that

$$|\langle f, v \rangle| = \lim_{n \to \infty} |\int_\Omega u_n \, \mathrm{div} \, v \, dx|$$

$$= \lim_{n \to \infty} |\int_\Omega \nabla u_n \cdot v \, dx|$$

$$\leq \lim_{n \to \infty} ||\nabla u_n||_p \left( \int_\Omega |v|^{p'} dx \right)^{1/p'}.$$

Since $\mathcal{D}(\Omega)$ is dense in $L^{p'}(\Omega)$, Proposition 3.2.3 implies the existence of a continuous extension of $f$ to $L^{p'}(\Omega; \mathbb{R}^N)$. By Riesz's representation theorem, there exists $g \in L^p(\Omega; \mathbb{R}^N)$ such that for every $v \in \mathcal{D}(\Omega; \mathbb{R}^N)$,

$$\int_\Omega g \cdot v \, dx = \langle f, v \rangle = \int_\Omega u \, \mathrm{div} \, v \, dx.$$

Hence $\nabla u = -g \in L^p(\Omega; \mathbb{R}^N)$. Choosing $v = |\nabla u|^{p-2} \nabla u$, we find that

$$\int_\Omega |\nabla u|^p dx = \int_\Omega \nabla u \cdot v \, dx \leq \lim_{n \to \infty} ||\nabla u_n||_p \left( \int_\Omega |v|^{p'} dx \right)^{1/p'}$$

$$= \lim_{n \to \infty} ||\nabla u_n||_p \left( \int_\Omega |\nabla u|^p dx \right)^{1-1/p}.$$

$\square$

Sobolev spaces are spaces of differentiable functions with integral norms. In order to define complete spaces, we use weak derivatives.

**Definition 6.1.8** Let $k \geq 1$ and $1 \leq p < \infty$. On the Sobolev space

$$W^{k,p}(\Omega) = \{u \in L^p(\Omega) : \text{ for every } |\alpha| \leq k, \, \partial^\alpha u \in L^p(\Omega)\},$$

we define the norm

$$||u||_{W^{k,p}(\Omega)} = ||u||_{k,p} = \left( \sum_{|\alpha| \leq k} \int_\Omega |\partial^\alpha u|^p dx \right)^{1/p}.$$

On the space $H^k(\Omega) = W^{k,2}(\Omega)$, we define the scalar product

$$(u \mid v)_{H^k(\Omega)} = \sum_{|\alpha| \le k} (\partial^\alpha u \mid \partial^\alpha v)_{L^2(\Omega)}.$$

The Sobolev space $W_{\mathrm{loc}}^{k,p}(\Omega)$ is defined by

$$W_{\mathrm{loc}}^{k,p}(\Omega) = \{u \in L_{\mathrm{loc}}^p(\Omega) : \text{ for all } \omega \subset\subset \Omega, u\big|_\omega \in W^{k,p}(\omega)\}.$$

A sequence $(u_n)$ converges to $u$ in $W_{\mathrm{loc}}^{k,p}(\Omega)$ if for every $\omega \subset\subset \Omega$,

$$||u_n - u||_{W^{k,p}(\omega)} \to 0, \quad n \to \infty.$$

The space $W_0^{k,p}(\Omega)$ is the closure of $\mathcal{D}(\Omega)$ in $W^{k,p}(\Omega)$. We denote by $H_0^k(\Omega)$ the space $W_0^{k,2}(\Omega)$.

**Theorem 6.1.9** *Let* $k \ge 1$ *and* $1 \le p < \infty$. *Then the spaces* $W^{k,p}(\Omega)$ *and* $W_0^{k,p}(\Omega)$ *are complete and separable.*

**Proof** Let $M = \sum_{|\alpha| \le k} 1$. The space $L^p(\Omega; \mathbb{R}^M)$ with the norm

$$||(v_\alpha)||_p = \left( \sum_{|\alpha| \le k} \int_\Omega |v_\alpha|^p dx \right)^{1/p}$$

is complete and separable. The map

$$\Phi : W^{k,p}(\Omega) \to L^p(\Omega; \mathbb{R}^M) : u \mapsto (\partial^\alpha u)_{|\alpha| \le k}$$

is a linear isometry: $||\Phi(u)||_p = ||u||_{k,p}$. By the closing lemma, $\Phi(W^{k,p}(\Omega))$ is a closed subspace of $L^p(\Omega; \mathbb{R}^M)$. It follows that $W^{k,p}(\Omega)$ is complete and separable. Since $W_0^{k,p}(\Omega)$ is a closed subspace of $W^{k,p}(\Omega)$, it is also complete and separable.
□

**Theorem 6.1.10** *Let* $1 \le p < \infty$. *Then* $W_0^{1,p}(\mathbb{R}^N) = W^{1,p}(\mathbb{R}^N)$.

**Proof** It suffices to prove that $\mathcal{D}(\mathbb{R}^N)$ is dense in $W^{1,p}(\mathbb{R}^N)$. We use regularization and truncation.

*Regularization* Let $u \in W^{1,p}(\mathbb{R}^N)$ and define $u_n = \rho_n * u$. By Proposition 4.3.6, $u_n \in C^\infty(\mathbb{R}^N)$. Proposition 4.3.14 implies that in $L^p(\mathbb{R}^N)$,

$$u_n \to u, \quad \partial_k u_n = \rho_n * \partial_k u \to \partial_k u.$$

We conclude that $W^{1,p}(\mathbb{R}^N) \cap C^\infty(\mathbb{R}^N)$ is dense in $W^{1,p}(\mathbb{R}^N)$.

*Truncation*  Let $\theta \in C^\infty(\mathbb{R})$ be such that $0 \leq \theta \leq 1$ and

$$\theta(t) = 1, \quad t \leq 1,$$
$$= 0, \quad t \geq 2.$$

We define the sequence

$$\theta_n(x) = \theta(|x|/n).$$

Let $u \in W^{1,p}(\mathbb{R}^N) \cap C^\infty(\mathbb{R}^N)$. It is clear that $u_n = \theta_n u \in \mathcal{D}(\mathbb{R}^N)$. It follows easily from Lebesgue's dominated convergence theorem that $u_n \to u$ in $W^{1,p}(\mathbb{R}^N)$.   □

We extend some rules of differential calculus to weak derivatives.

**Proposition 6.1.11 (Change of Variables)** *Let $\Omega$ and $\omega$ be open subsets of $\mathbb{R}^N$, $G : \omega \to \Omega$ a diffeomorphism, and $u \in W^{1,1}_{loc}(\Omega)$. Then $u \circ G \in W^{1,1}_{loc}(\omega)$ and*

$$\frac{\partial}{\partial y_k}(u \circ G) = \sum_j \frac{\partial u}{\partial x_j} \circ G \, \frac{\partial G_j}{\partial y_k}.$$

**Proof** Let $v \in \mathcal{D}(\omega)$ and $u_n = \rho_n * u$. By Lemma 6.1.1, for $n$ large enough, we have

$$\int_\omega u_n \circ G(y) \, \frac{\partial v}{\partial y_k}(y)dy = -\int_\omega \sum_j \frac{\partial u_n}{\partial x_j} \circ G(y) \, \frac{\partial G_j}{\partial y_k}(y) \, v(y)dy. \qquad (*)$$

It follows from Theorem 2.4.5 with $H = G^{-1}$ that

$$\int_\Omega u_n(x) \frac{\partial v}{\partial y_k} \circ H(x)| \det H'(x)|dx$$

$$= -\int_\Omega \sum_j \frac{\partial u_n}{\partial x_j}(x) \frac{\partial G_j}{\partial y_k} \circ H(x) v \circ H(x)| \det H'(x)|dx. \qquad (**)$$

The regularization theorem implies that in $L^1_{loc}(\Omega)$,

$$u_n \to u, \quad \frac{\partial u_n}{\partial x_j} \to \frac{\partial u}{\partial x_j}.$$

Taking the limit, it is permitted to replace $u_n$ by $u$ in $(**)$. But then it is also permitted to replace $u_n$ by $u$ in $(*)$, and the proof is complete.   □

**Proposition 6.1.12 (Derivative of a Product)** *Let $u \in W^{1,1}_{loc}(\Omega)$ and $f \in C^1(\Omega)$. Then $fu \in W^{1,1}_{loc}(\Omega)$ and*

$$\partial_k(fu) = f\partial_k u + (\partial_k f)u.$$

**Proof** Let $u_n = \rho_n * u$, so that by the classical rule of derivative of a product,

$$\partial_k(fu_n) = (\partial_k f)u_n + f\partial_k u_n.$$

It follows from the regularization theorem that

$$fu_n \to fu, \quad \partial_k(fu_n) \to (\partial_k f)u + f\partial_k u$$

in $L^1_{\text{loc}}(\Omega)$. We conclude by invoking the closing lemma. □

**Proposition 6.1.13 (Derivative of the Composition of Functions)** *Let* $u \in W^{1,1}_{\text{loc}}(\Omega)$, *and let* $f \in C^1(\mathbb{R})$ *be such that* $c = \sup_{\mathbb{R}}|f'| < \infty$. *Then* $f \circ u \in W^{1,1}_{\text{loc}}(\Omega)$
*and*

$$\partial_k(f \circ u) = f' \circ u \, \partial_k u.$$

**Proof** We define $u_n = \rho_n * u$, so that by the classical rule,

$$\partial_k(f \circ u_n) = f' \circ u_n \, \partial_k u_n.$$

We choose $\omega \subset\subset \Omega$. By the regularization theorem, we have in $L^1(\omega)$,

$$u_n \to u, \quad \partial_k u_n \to \partial_k u.$$

By Proposition 4.2.10, taking if necessary a subsequence, we can assume that $u_n \to u$ almost everywhere on $\omega$. We obtain

$$\int_\omega |f \circ u_n - f \circ u| dx \leq c \int_\omega |u_n - u| dx \to 0,$$

$$\int_\omega |f' \circ u_n \, \partial_k u_n - f' \circ u \, \partial_k u| dx \leq c \int_\omega |\partial_k u_n - \partial_k u| dx + \int_\omega |f' \circ u_n - f' \circ u| \, |\partial_k u| dx \to 0.$$

Hence in $L^1(\omega)$,

$$f \circ u_n \to f \circ u, \quad f' \circ u_n \, \partial_k u_n \to f' \circ u \, \partial_k u.$$

Since $\omega \subset\subset \Omega$ is arbitrary, we conclude the proof by invoking the closing lemma.

□

On $\mathbb{R}$, we define

$$\mathrm{sgn}(t) = t/|t|, \quad t \neq 0$$
$$= 0, \qquad t = 0.$$

**Corollary 6.1.14** *Let* $g : \mathbb{R} \to \mathbb{R}$ *be such that* $c = \sup_{\mathbb{R}} |g| < \infty$ *and, for some sequence* $(g_n) \subset C(\mathbb{R})$, $g(t) = \lim_{n \to \infty} g_n(t)$ *everywhere on* $\mathbb{R}$. *Define*

$$f(t) = \int_0^t g(s)ds.$$

*Then, for every* $u \in W^{1,1}_{loc}(\Omega)$, $f \circ u \in W^{1,1}_{loc}(\Omega)$ *and*

$$\nabla(f \circ u) = (g \circ u)\nabla u.$$

*In particular* $u^+, u^-, |u| \in W^{1,1}_{loc}(\Omega)$ *and*

$$\nabla u^+ = \chi_{\{u>0\}}\nabla u, \nabla u^- = -\chi_{\{u<0\}}\nabla u, \chi_{\{u=0\}}\nabla u = 0, \nabla|u| = (\mathrm{sgn}\, u)\nabla u.$$

**Proof** We can assume that $\sup_n \sup_{\mathbb{R}} |g_n| \leq c$. We define $f_n(t) = \int_0^t g_n(s)ds$. The preceding proposition implies that

$$\nabla(f_n \circ u) = (g_n \circ u)\nabla u.$$

Since, in $L^1_{loc}(\Omega)$, by Lebesgue's dominated convergence theorem,

$$f_n \circ u \to f \circ u, (g_n \circ u)\nabla u \to (g \circ u)\nabla u,$$

the closing lemma implies that

$$\nabla(f \circ u) = (g \circ u)\nabla u.$$

$$\square$$

**Corollary 6.1.15** *Let* $1 \leq p < \infty$ *and let* $u \in W^{1,p}(\Omega) \cap C(\overline{\Omega})$ *be such that* $u = 0$ *on* $\partial\Omega$. *Then* $u \in W^{1,p}_0(\Omega)$.

**Proof** It is easy to prove by regularization that $W^{1,p}(\Omega) \cap \mathcal{K}(\Omega) \subset W^{1,p}_0(\Omega)$.
Assume that spt $u$ is bounded. Let $f \in C^1(\mathbb{R})$ be such that $|f(t)| \leq |t|$ on $\mathbb{R}$,

$$f(t) = 0, \quad |t| \leq 1,$$
$$= t, \quad |t| \geq 2.$$

Define $u_n = f(n\,u)/n$. Then $u_n \in \mathcal{K}(\Omega)$, and by the preceding proposition, $u_n \in W^{1,p}(\Omega)$. By Lebesgue's dominated convergence theorem, $u_n \to u$ in $W^{1,p}(\Omega)$, so that $u \in W_0^{1,p}(\Omega)$.

If spt $u$ is unbounded, we define $u_n = \theta_n u$ where $(\theta_n)$ is defined in the proof of Theorem 6.1.10. Then spt $u_n$ is bounded. By Lebesgue's dominated convergence theorem, $u_n \to u$ in $W^{1,p}(\Omega)$, so that $u \in W_0^{1,p}(\Omega)$.                                 □

**Proposition 6.1.16** *Let $\Omega$ be an open subset of $\mathbb{R}^N$. Then there exist a sequence $(U_n)$ of open subsets of $\Omega$ and a sequence of functions $\psi_n \in \mathcal{D}(U_n)$ such that*

*(a) for every $n$, $U_n \subset\subset \Omega$ and $\psi_n \geq 0$;*

*(b) $\displaystyle\sum_{n=1}^{\infty} \psi_n = 1$ on $\Omega$;*

*(c) for every $\omega \subset\subset \Omega$ there exists $m_\omega$ such that for $n > m_\omega$ we have $U_n \cap \omega = \phi$.*

***Proof*** Let us define $\omega_{-1} = \omega_0 = \phi$, and for $n \geq 1$,

$$\omega_n = \{x \in \Omega : d(x, \partial\Omega) > 1/n \text{ and } |x| < n\},$$
$$K_n = \overline{\omega_n} \setminus \omega_{n-1},$$
$$U_n = \omega_{n+1} \setminus \overline{\omega_{n-2}}.$$

The theorem of partitions of unity implies the existence of $\varphi_n \in \mathcal{D}(U_n)$ such that $0 \leq \varphi_n \leq 1$ and $\varphi_n = 1$ on $K_n$. It suffices then to define

$$\psi_n = \varphi_n \Big/ \sum_{j=1}^{\infty} \varphi_j.$$                                 □

**Theorem 6.1.17 (Hajłasz)** *Let $1 \leq p < \infty$, $u \in W_{\text{loc}}^{1,p}(\Omega)$, and $\varepsilon > 0$. Then there exists $v \in C^{\infty}(\Omega)$ such that*

*(a) $v - u \in W_0^{1,p}(\Omega)$;*
*(b) $\|v - u\|_{W^{1,p}(\Omega)} < \varepsilon$.*

***Proof*** Let $(U_n)$ and $(\psi_n)$ be given by the preceding proposition. For every $n \geq 1$, there exists $k_n$ such that

$$v_n = \rho_{k_n} * (\psi_n u) \in \mathcal{D}(U_n)$$

and

$$\|v_n - \psi_n u\|_{1,p} < \varepsilon/2^n.$$

By Proposition 3.1.6, $\sum\limits_{n=1}^{\infty}(v_n - \psi_n u)$ converges to $w$ in $W_0^{1,p}(\Omega)$. On the other hand, we have on $\omega \subset\subset \Omega$ that

$$\sum_{n=1}^{\infty} v_n = \sum_{n=1}^{m_\omega} v_n \in C^\infty(\omega), \quad \sum_{n=1}^{\infty} \psi_n u = u.$$

Setting $v = \sum\limits_{n=1}^{\infty} v_n$, we conclude that

$$\|v - u\|_{1,p} = \|w\|_{1,p} \leq \sum_{n=1}^{\infty} \|v_n - \psi_n u\|_{1,p} < \varepsilon. \qquad \square$$

**Corollary 6.1.18 (Deny–Lions)** *Let* $1 \leq p < \infty$. *Then* $C^\infty(\Omega) \cap W^{1,p}(\Omega)$ *is dense in* $W^{1,p}(\Omega)$.

## 6.2 Cylindrical Domains

Let $U$ be an open subset of $\mathbb{R}^{N-1}$ and $0 < r \leq \infty$. Define

$$\Omega = U \times\ ]-r, r[, \quad \Omega_+ = U \times\ ]0, r[.$$

The extension by reflection of a function in $W^{1,p}(\Omega_+)$ is a function in $W^{1,p}(\Omega)$. For every $u : \Omega_+ \to \mathbb{R}$, we define on $\Omega$:

$$\rho u(x', x_N) = u\left(x', |x_N|\right), \quad \sigma u(x', x_N) = (\operatorname{sgn} x_N) u\left(x', |x_N|\right).$$

**Lemma 6.2.1 (Extension by Reflection)** *Let* $1 \leq p < \infty$ *and* $u \in W^{1,p}(\Omega_+)$. *Then* $\rho u \in W^{1,p}(\Omega)$, $\partial_k(\rho u) = \rho(\partial_k u)$, $1 \leq k \leq N - 1$, *and* $\partial_N(\rho u) = \sigma(\partial_N u)$, *so that*

$$\|\rho u\|_{L^p(\Omega)} = 2^{1/p}\|u\|_{L^p(\Omega_+)}, \quad \|\rho u\|_{W^{1,p}(\Omega)} = 2^{1/p}\|u\|_{W^{1,p}(\Omega_+)}.$$

**Proof** Let $v \in \mathcal{D}(\Omega)$. Then by a change of variables,

$$\int_\Omega (\rho u)\partial_N v \, dx = \int_{\Omega_+} u\, \partial_N w \, dx, \qquad (*)$$

where

$$w(x', x_N) = v(x', x_N) - v(x', -x_N).$$

A truncation argument will be used. Let $\eta \in C^\infty(\mathbb{R})$ be such that

$$\eta(t) = 0, \quad t < 1/2,$$
$$= 1, \quad t > 1,$$

and define $\eta_n$ on $\Omega_+$ by

$$\eta_n(x) = \eta(n x_N).$$

The definition of weak derivative ensures that

$$\int_{\Omega_+} u \, \partial_N (\eta_n w) dx = - \int_{\Omega_+} (\partial_N u) \eta_n w \, dx, \qquad (\ast\ast)$$

where

$$\partial_N (\eta_n w) = \eta_n \partial_N w + n\eta'(n x_N) w.$$

Since $w(x', 0) = 0$, $w(x', x_N) = h(x', x_N) x_N$, where

$$h(x', x_N) = \int_0^1 \partial_N w(x', t x_N) dt.$$

Lebesgue's dominated convergence theorem implies that

$$\left| \int_{\Omega_+} n \, \eta'(n x_N) w \, u \, dx \right| = \left| \int_{U \times \, ]0, 1/n[} n \, \eta'(n x_N) h \, x_N u \, dx \right|$$

$$\leq \|\eta'\|_\infty \int_{U \times \, ]0, 1/n[} |hu| dx \to 0, \quad n \to \infty.$$

Taking the limit in $(\ast\ast)$, we obtain

$$\int_{\Omega_+} u \, \partial_N w \, dx = - \int_{\Omega_+} (\partial_N u) w \, dx = - \int_\Omega \sigma(\partial_N u) v \, dx.$$

It follows from $(\ast)$ that

$$\int_\Omega (\rho u) \partial_N v \, dx = - \int_\Omega \sigma(\partial_N u) v \, dx.$$

Since $v \in \mathcal{D}(\Omega)$ is arbitrary, $\partial_N(\rho u) = \sigma(\partial_N u)$. By a similar but simpler argument, $\partial_k(\rho u) = \rho(\partial_k u)$, $1 \leq k \leq N-1$. $\qquad\square$

It makes no sense to define an $L^p$ function on a set of measure zero. We will define the trace of a $W^{1,p}$ function on the boundary of a smooth domain. We first consider the case of $\mathbb{R}^N_+$.

*Notation*  We define

$$\mathcal{D}(\overline{\Omega}) = \{u|_\Omega : u \in \mathcal{D}(\mathbb{R}^N)\},$$

$$\mathbb{R}^N_+ = \{(x', x_N) : x' \in \mathbb{R}^{N-1}, x_N > 0\}.$$

**Lemma 6.2.2 (Trace Inequality)**  *Let* $1 \leq p < \infty$. *Then for every* $u \in \mathcal{D}(\overline{\mathbb{R}^N_+})$,

$$\int_{\mathbb{R}^{N-1}} \left|u(x', 0)\right|^p dx' \leq p\|u\|_{L^p(\mathbb{R}^N_+)}^{p-1}\|\partial_N u\|_{L^p_{(\mathbb{R}^N_+)}}.$$

***Proof***  The fundamental theorem of calculus implies that for all $x' \in \mathbb{R}^{N-1}$,

$$\left|u(x', 0)\right|^p \leq p\int_0^\infty \left|u(x', x_N)\right|^{p-1}\left|\partial_N u(x', x_N)\right| dx_N.$$

When $1 < p < \infty$, using Fubini's theorem and Hölder's inequality, we obtain

$$\int_{\mathbb{R}^{N-1}} \left|u(x', 0)\right|^p dx' \leq p\int_{\mathbb{R}^N_+} |u|^{p-1}|\partial_N u|dx$$

$$\leq p\left(\int_{\mathbb{R}^N_+} |u|^{(p-1)p'}dx\right)^{1/p'}\left(\int_{\mathbb{R}^N_+} |\partial_N u|^p dx\right)^{1/p}$$

$$= p\left(\int_{\mathbb{R}^N_+} |u|^p dx\right)^{1-1/p}\left(\int_{\mathbb{R}^N_+} |\partial_N u|^p dx\right)^{1/p}.$$

The case $p = 1$ is similar. $\qquad\square$

**Proposition 6.2.3**  *Let* $1 \leq p < \infty$. *Then there exists one and only one continuous linear mapping* $\gamma_0 : W^{1,p}(\mathbb{R}^N_+) \to L^p(\mathbb{R}^{N-1})$ *such that for every* $u \in \mathcal{D}(\overline{\mathbb{R}^N_+})$, $\gamma_0 u = u(., 0)$.

***Proof***  Let $u \in \mathcal{D}(\overline{\mathbb{R}^N_+})$ and define $\gamma_0 u = u(., 0)$. The preceding lemma implies that

$$\|\gamma_0 u\|_{L^p(\mathbb{R}^{N-1})} \leq p^{1/p}\|u\|_{W^{1,p}(\mathbb{R}^N_+)}.$$

The space $\mathcal{D}(\overline{\mathbb{R}^N_+})$ is dense in $W^{1,p}(\mathbb{R}^N_+)$ by Theorem 6.1.10 and Lemma 6.2.1. By Proposition 3.2.3, $\gamma_0$ has a unique continuous linear extension to $W^{1,p}(\mathbb{R}^N_+)$.  □

**Proposition 6.2.4 (Integration by Parts)** *Let* $1 \leq p < \infty$, $u \in W^{1,p}(\mathbb{R}^N_+)$, *and* $v \in \mathcal{D}(\overline{\mathbb{R}^N_+})$. *Then*

$$\int_{\mathbb{R}^N_+} v\,\partial_N u\,dx = -\int_{\mathbb{R}^N_+} (\partial_N v)u\,dx - \int_{\mathbb{R}^{N-1}} \gamma_0 v\,\gamma_0 u\,dx',$$

*and*

$$\int_{\mathbb{R}^N_+} v\partial_k u\,dx = -\int_{\mathbb{R}^N_+} (\partial_k v)u\,dx, \quad 1 \leq k \leq N-1.$$

***Proof*** Assume, moreover, that $u \in \mathcal{D}(\overline{\mathbb{R}^N_+})$. Integrating by parts, we obtain for all $x' \in \mathbb{R}^{N-1}$,

$$\int_0^\infty v(x',x_N)\partial_N u(x',x_N)dx_N = -\int_0^\infty \partial_N v(x',x_N)u(x',x_N)dx_N - v(x',0)u(x',0).$$

Fubini's theorem implies that

$$\int_{\mathbb{R}^N_+} v\,\partial_N u\,dx = -\int_{\mathbb{R}^N_+} \partial_N vu\,dx - \int_{\mathbb{R}^{N-1}} v(x',0)u(x',0)dx'.$$

Let $u \in W^{1,p}(\mathbb{R}^N_+)$. Since $\mathcal{D}(\overline{\mathbb{R}^N_+})$ is dense in $W^{1,p}(\mathbb{R}^N_+)$, there exists a sequence $(u_n) \subset \mathcal{D}(\overline{\mathbb{R}^N_+})$ such that $u_n \to u$ in $W^{1,p}(\mathbb{R}^N_+)$. By the preceding lemma, $\gamma_0 u_n \to \gamma_0 u$ in $L^p(\mathbb{R}^{N-1})$. It is easy to finish the proof.

The proof of the last formulas is similar.  □

*Notation* For every $u : \mathbb{R}^N_+ \to \mathbb{R}$, we define $\overline{u}$ on $\mathbb{R}^N$ by

$$\overline{u}(x',x_N) = u(x',x_N), \quad x_N > 0,$$
$$= 0, \quad\quad\quad\quad\; x_N \leq 0.$$

**Proposition 6.2.5** *Let* $1 \leq p < \infty$ *and* $u \in W^{1,p}(\mathbb{R}^N_+)$. *The following properties are equivalent:*

(a) $u \in W_0^{1,p}(\mathbb{R}^N_+)$;
(b) $\gamma_0 u = 0$;
(c) $\overline{u} \in W^{1,p}(\mathbb{R}^N)$ *and* $\partial_k \overline{u} = \overline{\partial_k u}$, $1 \leq k \leq N$.

**Proof** If $u \in W_0^{1,p}(\mathbb{R}_+^N)$, there exists $(u_n) \subset \mathcal{D}(\mathbb{R}_+^N)$ such that $u_n \to u$ in $W^{1,p}(\mathbb{R}_+^N)$. Hence $\gamma_0 u_n = 0$ and $\gamma_0 u_n \to \gamma_0 u$ in $L^p(\mathbb{R}^{N-1})$, so that $\gamma_0 u = 0$.

If $\gamma_0 u = 0$, it follows from the preceding proposition that for every $v \in \mathcal{D}(\mathbb{R}^N)$,

$$\int_{\mathbb{R}^N} v \, \overline{\partial_k u} \, dx = - \int_{\mathbb{R}^N} \partial_k v \, \bar{u} \, dx, \quad 1 \le k \le N.$$

We conclude that (c) is satisfied.

Assume that (c) is satisfied. We define $u_n = \theta_n \bar{u}$, where $(\theta_n)$ is defined in the proof of Theorem 6.1.10. It is clear that $u_n \to \bar{u}$ in $W^{1,p}(\mathbb{R}^N)$ and spt $u_n \subset B[0, 2n] \cap \overline{\mathbb{R}_+^N}$.

We can assume that spt $u_n$ is a compact subset of $\overline{\mathbb{R}_+^N}$. We define $y_n = (0, \ldots, 0, 1/n)$ and $v_n = \tau_{y_n} \bar{u}$. Since $\partial_k v_n = \tau_{y_n} \partial_k \bar{u}$, the lemma of continuity of translations implies that $u_n \to u$ in $W^{1,p}(\mathbb{R}_+^N)$.

We can assume that spt $u$ is a compact subset of $\mathbb{R}_+^N$. For $n$ large enough, $\rho_n * u \in \mathcal{D}(\mathbb{R}_+^N)$. Since $\rho_n * u \to u$ is in $W^{1,p}(\mathbb{R}^N)$, we conclude that $u \in W_0^{1,p}(\mathbb{R}^N)$.   $\square$

## 6.3   Smooth Domains

In this section we consider an open subset $\Omega = \{\varphi < 0\}$ of $\mathbb{R}^N$ of class $C^1$ with a bounded boundary $\Gamma$. We use the notations of Definition 9.4.1.

Let $\gamma \in \Gamma$. Since $\nabla \varphi(\gamma) \ne 0$, we can assume that, after a permutation of variables, $\partial_N \varphi(\gamma) \ne 0$. By Theorem 9.1.1 there exist $r > 0$, $R > 0$, and

$$\beta \in C^1\big(B(\gamma', R) \times ] - r, r[\big)$$

such that, for $|x' - \gamma'| < R$ and $|t| < r$, we have

$$\varphi(x', x_N) = t \quad \Leftrightarrow \quad x_N = \beta(x', t)$$

and the set

$$U_\gamma = \Big\{ (x', \beta(x', t)) : |x' - \gamma'| < R, |t| < r \Big\}$$

is an open neighborhood of $\gamma$. Moreover

$$\Omega \cap U_\gamma = \Big\{ (x', \beta(x', t)) : |x' - \gamma'| < R, -r < t < 0 \Big\}$$

and

$$\Gamma \cap U_\gamma = \left\{ (x', \beta(x', 0)): |x' - \gamma'| < R \right\}.$$

The Borel–Lebesgue theorem implies the existence of a finite covering $U_1, \ldots, U_k$ of $\Gamma$ by open subsets satisfying the above properties. There exists a partition of unity $\psi_1, \ldots, \psi_k$ subordinate to this covering.

**Theorem 6.3.1 (Extension Theorem)** *Let* $1 \le p < \infty$ *and let* $\Omega$ *be an open subset of* $\mathbb{R}^N$ *of class* $C^1$ *with a bounded boundary or the product of* $N$ *open intervals. Then there exists a continuous linear mapping*

$$P : W^{1,p}(\Omega) \to W^{1,p}(\mathbb{R}^N)$$

*such that* $Pu\big|_\Omega = u.$

**Proof** Let $\Omega$ be an open subset of $\mathbb{R}^N$ of class $C^1$ with a bounded boundary, and let $u \in W^{1,p}(\Omega)$. Proposition 6.1.11 and Lemma 6.2.1 imply that

$$P_U u(x) = u(x', \beta(x', -|\varphi(x', x_N)|)) \in W^{1,p}(U).$$

Moreover,

$$\|P_U u\|_{W^{1,p}(U)} \le a_U \|u\|_{W^{1,p}(\Omega)}. \tag{$*$}$$

We define $\psi_0 = 1 - \displaystyle\sum_{j=1}^{k} \psi_j,$

$$\begin{aligned} u_0 &= \psi_0 u, & x \in \Omega, \\ &= 0, & x \in \mathbb{R}^N \setminus \Omega, \end{aligned}$$

and for $1 \le j \le k$,

$$\begin{aligned} u_j &= P_{U_j}(\psi_j u), & x \in U_j, \\ &= 0, & x \in \mathbb{R}^N \setminus U_j. \end{aligned}$$

Formula $(*)$ and Proposition 6.1.12 ensure that for $0 \le j \le k$,

$$\|u_j\|_{W^{1,p}(\mathbb{R}^N)} \le b_j \|u\|_{W^{1,p}(\Omega)}.$$

(The support of $\nabla \psi_0$ is compact!) Hence

$$Pu = \sum_{j=0}^{k} u_j \in W^{1,p}(\mathbb{R}^N), \quad \|Pu\|_{W^{1,p}(\mathbb{R}^N)} \le c \|u\|_{W^{1,p}(\Omega)},$$

and for all $x \in \Omega$,

$$Pu(x) = \sum_{j=0}^{k} \psi_j(x)u(x) = u(x).$$

If $\Omega$ is the product of $N$ open intervals, it suffices to use a finite number of extensions by reflections and a truncation.  □

**Theorem 6.3.2 (Density Theorem in Sobolev Spaces)** *Let $1 \le p < \infty$ and let $\Omega$ be an open subset of $\mathbb{R}^N$ of class $C^1$ with a bounded boundary or the product of $N$ open intervals. Then the space $\mathcal{D}(\bar{\Omega})$ is dense in $W^{1,p}(\Omega)$.*

**Proof** Let $u \in W^{1,p}(\Omega)$. Theorem 6.1.10 implies the existence of a sequence $(v_n) \subset \mathcal{D}(\mathbb{R}^N)$ converging to $Pu$ in $W^{1,p}(\mathbb{R}^N)$. Hence $u_n = v_n|_\Omega$ converges to $u$ in $W^{1,p}(\Omega)$.  □

**Theorem 6.3.3 (Trace Inequality)** *Let $\Omega$ be an open subset of $\mathbb{R}^N$ of class $C^1$ with a bounded boundary $\Gamma$. Then there exist $a > 0$ and $b > 0$ such that, for $1 \le p < \infty$ and for every $u \in \mathcal{D}(\bar{\Omega})$,*

$$\int_\Gamma |u|^p d\gamma \le a\|u\|_{L^p(\Omega)}^p + bp\|u\|_{L^p(\Omega)}^{p-1}\|\nabla u\|_{L^p(\Omega)}.$$

**Proof** Let $1 < p < \infty$, $u \in \mathcal{D}(\bar{\Omega})$, and $v \in C^\infty(\mathbb{R}^N; \mathbb{R}^N)$. Since

$$\operatorname{div}|u|^p v = |u|^p \operatorname{div} v + pu|u|^{p-2}\nabla u \cdot v,$$

the divergence theorem implies that

$$\int_\Gamma |u|^p v \cdot n d\gamma = \int_\Omega \left[ |u|^p \operatorname{div} v + pu|u|^{p-2}\nabla u \cdot v \right] dx.$$

Assume that $1 \le v \cdot n$ on $\Gamma$. Using Hölder's inequality, we obtain that, for $1 < p < \infty$,

$$\int_\Gamma |u|^p d\gamma \le \int_\Gamma |u|^p v \cdot n d\gamma \le a \int_\Omega |u|^p dx + bp \int_\Omega |u|^{p-1}|\nabla u|dx$$

$$\le a \int_\Omega |u|^p dx + bp \left( \int_\Omega |u|^{(p-1)p'} dx \right)^{1/p'} \left( \int_\Omega |\nabla u|^p dx \right)^{1/p}$$

$$= a \int_\Omega |u|^p dx + bp \left( \int_\Omega |u|^p dx \right)^{1-1/p} \left( \int_\Omega |\nabla u|^p dx \right)^{1/p},$$

where $a = \|\operatorname{div} v\|_\infty$ and $b = \|v\|_\infty$.

When $p \downarrow 1$, it follows from Lebesgue's dominated convergence theorem that

$$\int_\Gamma |u| d\gamma \le a \int_\Omega |u| dx + b \int_\Omega |\nabla u| dx.$$

Let us construct an admissible vector field $v$. Let $U = \{x \in \mathbb{R}^N : \nabla\varphi(x) \ne 0\}$. The theorem of partition of unity implies the existence of $\psi \in \mathcal{D}(U)$ such that $\psi = 1$ on $\Gamma$. We define the vector field $w$ by

$$w(x) = \psi(x) \frac{\nabla\varphi(x)}{|\nabla\varphi(x)|}, \quad x \in U$$

$$= 0, \qquad\qquad x \in \mathbb{R}^N \setminus U.$$

For $n$ large enough, the $C^\infty$ vector field $v = 2\rho_n * w$ is such that $1 \le v \cdot n$ on $\Gamma$. $\qquad\square$

**Theorem 6.3.4** *Under the assumptions of Theorem 6.3.3, there exists one and only one continuous linear mapping*

$$\gamma : W^{1,p}(\Omega) \to L^p(\Gamma)$$

*such that for all $u \in \mathcal{D}(\bar{\Omega})$, $\gamma_0 u = u\big|_\Gamma$.*

**Proof** It suffices to use the trace inequality, Proposition 3.2.3, and the density theorem in Sobolev spaces. $\qquad\square$

**Theorem 6.3.5 (Divergence Theorem)** *Let $\Omega$ be an open subset of $\mathbb{R}^N$ of class $C^1$ with a bounded boundary $\Gamma$ and $v \in W^{1,1}(\Omega; \mathbb{R}^N)$. Then*

$$\int_\Omega \operatorname{div} v dx = \int_\Gamma \gamma_0 v \cdot n d\gamma.$$

**Proof** When $v \in \mathcal{D}(\bar{\Omega}; \mathbb{R}^N)$, the proof is given in Section 9.4. In the general case, it suffices to use the density theorem in Sobolev spaces and the trace theorem. $\qquad\square$

## 6.4  Embeddings

Let $1 \le p, q < \infty$. If there exists $c > 0$ such that for every $u \in \mathcal{D}(\mathbb{R}^N)$,

$$\|u\|_{L^q(\mathbb{R}^N)} \le c\|\nabla u\|_{L^p(\mathbb{R}^N)},$$

then necessarily

$$q = p^* = Np/(N - p).$$

Indeed, replacing $u(x)$ by $u_\lambda(x) = u(\lambda x)$, $\lambda > 0$, we find that

$$\|u\|_{L^q(\mathbb{R}^N)} \le c\lambda^{\left(1 + \frac{N}{q} - \frac{N}{p}\right)} \|\nabla u\|_{L^p(\mathbb{R}^N)},$$

so that $q = p^*$.

We define for $1 \le j \le N$ and $x \in \mathbb{R}^N$,

$$\widehat{x}_j = (x_1, \ldots, x_{j-1}, x_{j+1}, \ldots, x_N).$$

**Lemma 6.4.1 (Gagliardo's Inequality)** *Let* $N \ge 2$ *and* $f_1, \ldots, f_N \in L^{N-1}(\mathbb{R}^{N-1})$. *Then* $f(x) = \prod_{j=1}^{N} f_j(\widehat{x}_j) \in L^1(\mathbb{R}^N)$ *and*

$$\|f\|_{L^1(\mathbb{R}^N)} \le \prod_{j=1}^{N} \|f_j\|_{L^{N-1}(\mathbb{R}^{N-1})}.$$

***Proof*** We use induction. When $N = 2$, the inequality is clear. Assume that the inequality holds for $N \ge 2$. Let $f_1, \ldots, f_{N+1} \in L^N(\mathbb{R}^N)$ and

$$f(x, x_{N+1}) = \prod_{j=1}^{N} f_j(\widehat{x}_j, x_{N+1}) f_{N+1}(x).$$

It follows from Hölder's inequality that for almost every $x_{N+1} \in \mathbb{R}$,

$$\int_{\mathbb{R}^N} |f(x, x_{N+1})| dx \le \left[ \int_{\mathbb{R}^N} \prod_{j=1}^{N} |f_j(\widehat{x}_j, x_{N+1})|^{N'} dx \right]^{1/N'} \|f_{N+1}\|_{L^N(\mathbb{R}^N)}$$

$$\le \prod_{j=1}^{N} \left[ \int_{\mathbb{R}^{N-1}} |f_j(\widehat{x}_j, x_{N+1})|^N d\widehat{x}_j \right]^{1/N} \|f_{N+1}\|_{L^N(\mathbb{R}^N)}.$$

The generalized Hölder inequality implies that

$$||f||_{L^1(\mathbb{R}^{N+1})} \leq \prod_{j=1}^{N} \left[ \int_{\mathbb{R}^N} |f_j(\widehat{x_j}, x_{N+1})|^N d\widehat{x_j} dx_{N+1} \right]^{1/N} ||f_{N+1}||_{L^N(\mathbb{R}^N)}$$

$$= \prod_{j=1}^{N+1} ||f_j||_{L^N(\mathbb{R}^N)}. \qquad \qquad \square$$

**Lemma 6.4.2 (Sobolev's Inequalities)** *Let* $1 \leq p < N$. *Then there exists a constant* $c = c(p, N)$ *such that for every* $u \in \mathcal{D}(\mathbb{R}^N)$,

$$||u||_{L^{p^*}(\mathbb{R}^N)} \leq c||\nabla u||_{L^p(\mathbb{R}^N)}.$$

**Proof** Let $u \in C^1(\mathbb{R}^N)$ be such that spt $u$ is compact. It follows from the fundamental theorem of calculus that for $1 \leq j \leq N$ and $x \in \mathbb{R}^N$,

$$|u(x)| \leq \frac{1}{2} \int_{\mathbb{R}} |\partial_j u(x)| dx_j.$$

By the preceding lemma,

$$\int_{\mathbb{R}^N} |u(x)|^{N/(N-1)} dx \leq \prod_{j=1}^{N} \left[ \frac{1}{2} \int_{\mathbb{R}^N} |\partial_j u(x)| dx \right]^{1/(N-1)}.$$

Hence we obtain

$$||u||_{N/(N-1)} \leq \frac{1}{2} \prod_{j=1}^{N} ||\partial_j u||_1^{1/N} \leq c_N ||\nabla u||_1.$$

For $p > 1$, we define $q = (N-1)p^*/N > 1$. Let $u \in \mathcal{D}(\mathbb{R}^N)$. The preceding inequality applied to $|u|^q$ and Hölder's inequality imply that

$$\left( \int |u|^{p^*} dx \right)^{\frac{N-1}{N}} \leq q \, c_N \int_{\mathbb{R}^N} |u|^{q-1} |\nabla u| dx$$

$$\leq q \, c_N \left( \int_{\mathbb{R}^N} |u|^{(q-1)p'} dx \right)^{1/p'} \left( \int_{\mathbb{R}^N} |\nabla u|^p dx \right)^{1/p}.$$

It is easy to conclude the proof. $\qquad \qquad \square$

**Lemma 6.4.3 (Morrey's Inequalities)** *Let* $N < p < \infty$ *and* $\lambda = 1 - N/p$. *Then there exists a constant* $c = c(p, N)$ *such that for every* $u \in \mathcal{D}(\mathbb{R}^N)$ *and every* $x, y \in \mathbb{R}^N$,

$$\left|u(x) - u(y)\right| \leq c|x - y|^\lambda \|\nabla u\|_{L^P(\mathbb{R}^N)},$$

$$\|u\|_\infty \leq c\|u\|_{W^{1,P}(\mathbb{R}^N)}.$$

**Proof** Let $u \in \mathcal{D}(\mathbb{R}^N)$, and let us define $B = B(a, r), a \in \mathbb{R}^N, r > 0$, and

$$\fint u = \frac{1}{m(B)} \int_B u \, dx.$$

We assume that $0 \in \bar{B}$. It follows from the fundamental theorem of calculus and Fubini's theorem that

$$\left|\fint u - u(0)\right| \leq \frac{1}{m(B)} \int_B |u(x) - u(0)| dx$$

$$\leq \frac{1}{m(B)} \int_B dx \int_0^1 \left|\nabla u(tx)\right| |x| dt$$

$$\leq \frac{2r}{m(B)} \int_0^1 dt \int_B \left|\nabla u(tx)\right| dx$$

$$= \frac{2r}{m(B)} \int_0^1 \frac{dt}{t^N} \int_{B(ta, tr)} \left|\nabla u(y)\right| dy.$$

Hölder's inequality implies that

$$\left|\fint u - u(0)\right| \leq \frac{2r}{m(B)} \int_0^1 m\big(B(ta, tr)\big)^{1/p'} \frac{dt}{t^N} \|\nabla u\|_{L^P(B)} = \frac{2}{\lambda V_N^{1/p}} r^\lambda \|\nabla u\|_{L^P(B)}.$$

After a translation, we obtain that, for every $x \in B[a, r]$,

$$\left|\fint u - u(x)\right| \leq c_\lambda r^\lambda \|\nabla u\|_{L^P(B)}.$$

Let $x \in \mathbb{R}^N$. Choosing $a = x$ and $r = 1$, we find

$$|u(x)| \leq \left|\fint u\right| + c_\lambda \|\nabla u\|_{L^P(B)} \leq c\big(\|u\|_{L^P(B)} + \|\nabla u\|_{L^P(B)}\big).$$

Let $x, y \in \mathbb{R}^N$. Choosing $a = (x + y)/2$ and $r = |x - y|/2$, we obtain

$$|u(x) - u(y)| \leq 2^{1-\lambda} c_\lambda |x - y|^\lambda \|\nabla u\|_{L^P(B)}. \qquad \square$$

*Notation* We define

$$C_0(\overline{\Omega}) = \{u\big|_{\Omega} : u \in C_0(\mathbb{R}^N)\}.$$

**Theorem 6.4.4 (Sobolev's Embedding Theorem, 1936–1938)** *Let $\Omega$ be an open subset of $\mathbb{R}^N$ of class $C^1$ with a bounded boundary or the product of $N$ open intervals.*

(a) *If $1 \le p < N$ and if $p \le q \le p^*$, then $W^{1,p}(\Omega) \subset L^q(\Omega)$, and the canonical injection is continuous.*

(b) *If $N < p < \infty$ and $\lambda = 1 - N/p$, then $W^{1,p}(\Omega) \subset C_0(\overline{\Omega})$, the canonical injection is continuous, and there exists $c = c(p, \Omega)$ such that for every $u \in W^{1,p}(\Omega)$ and all $x, y \in \Omega$,*

$$\big|u(x) - u(y)\big| \le c\|u\|_{W^{1,p}(\Omega)}|x - y|^{\lambda}.$$

**Proof** Let $1 \le p < N$ and $u \in W^{1,p}(\mathbb{R}^N)$. By Theorem 6.1.10, there exists a sequence $(u_n) \subset \mathcal{D}(\mathbb{R}^N)$ such that $u_n \to u$ in $W^{1,p}(\mathbb{R}^N)$.

We can assume that $u_n \to u$ almost everywhere on $\mathbb{R}^N$. It follows from Fatou's lemma and Sobolev's inequality that

$$\|u\|_{L^{p^*}(\mathbb{R}^N)} \le \varliminf_{n\to\infty} \|u_n\|_{L^{p^*}(\mathbb{R}^N)} \le c \lim_{n\to\infty} \|\nabla u_n\|_{L^p(\mathbb{R}^N)} = c\|\nabla u\|_{L^p(\mathbb{R}^N)}.$$

Let $P$ be the extension operator corresponding to $\Omega$ and $v \in W^{1,p}(\Omega)$. We have

$$\|v\|_{L^{p^*}(\Omega)} \le \|Pv\|_{L^{p^*}(\mathbb{R}^N)} \le c\|\nabla Pv\|_{L^p(\mathbb{R}^N)} \le c_1\|v\|_{W^{1,p}(\Omega)}.$$

If $p \le q \le p^*$, we define $0 \le \lambda \le 1$ by

$$\frac{1}{q} = \frac{1-\lambda}{p} + \frac{\lambda}{p^*},$$

and we infer from the interpolation inequality that

$$\|v\|_{L^q(\Omega)} \le \|v\|_{L^p(\Omega)}^{1-\lambda}\|v\|_{L^{p^*}(\Omega)}^{\lambda} \le c_1^{\lambda}\|v\|_{W^{1,p}(\Omega)}.$$

The case $p > N$ follows from Morrey's inequalities.                                      $\square$

**Lemma 6.4.5** *Let $\Omega$ be an open subset of $\mathbb{R}^N$ such that $m(\Omega) < +\infty$, and let $1 \le p \le r < +\infty$. Assume that $X$ is a closed subspace of $W^{1,p}(\Omega)$ such that $X \subset L^r(\Omega)$. Then, for every $1 \le q < r$, $X \subset L^q(\Omega)$ and the canonical injection is compact.*

***Proof*** The closed graph theorem implies the existence of $c > 0$ such that, for every $u \in X$,

$$\|u\|_{L^r(\Omega)} \leq c\|u\|_{W^{1,p}(\Omega)}.$$

Our goal is to prove that

$$S = \{u \in X : \|u\|_{W^{1,p}(\Omega)} \leq 1\}$$

is precompact in $L^q(\Omega)$ for $1 \leq q < r$. Let $1/q = 1 - \lambda + \lambda/r$. By the interpolation inequality, for every $u \in S$,

$$\|u\|_{L^q(\Omega)} \leq \|u\|_{L^r(\Omega)}^{\lambda} \|u\|_{L^1(\Omega)}^{1-\lambda} \leq c^{\lambda} \|u\|_{L^1(\Omega)}^{1-\lambda}.$$

Hence it suffices to prove that $S$ is precompact in $L^1(\Omega)$.

Let us verify that $S$ satisfies the assumptions of M. Riesz's theorem in $L^1(\Omega)$:

(a) It follows from Hölder's inequality that, for every $u \in S$,

$$\|u\|_{L^1(\Omega)} \leq \|u\|_{L^r(\Omega)} m(\Omega)^{1-1/r} \leq cm(\Omega)^{1-1/r}.$$

(b) Similarly, we have that, for every $u \in S$,

$$\int_{\Omega \setminus \omega_k} |u| dx \leq \|u\|_{L^r(\Omega)} m(\Omega \setminus \omega_k)^{1-1/r} \leq cm(\Omega \setminus \omega_k)^{1-1/r}$$

where

$$\omega_k = \{x \in \Omega : d(x, \partial\Omega) > 1/k\}.$$

Lebesgue's dominated convergence theorem implies that

$$\lim_{k \to \infty} m(\Omega \setminus \omega_k) = 0.$$

(c) Let $\omega \subset\subset \Omega$. Assume that $|y| < d(\omega, \partial\Omega)$ and $u \in C^{\infty}(\Omega) \cap W^{1,p}(\Omega)$.

Since, by the fundamental theorem of calculus,

$$\left| \tau_y u(x) - u(x) \right| = \left| \int_0^1 y \cdot \nabla u(x - ty) dt \right| \leq |y| \int_0^1 \left| \nabla u(x - ty) \right| dt,$$

we obtain

$$\|\tau_y u - u\|_{L^1(\omega)} \le |y| \int_\omega dx \int_0^1 \left|\nabla u(x - ty)\right| dt$$

$$= |y| \int_0^1 dt \int_\omega \left|\nabla u(x - ty)\right| dx$$

$$= |y| \int_0^1 dt \int_{\omega - ty} \left|\nabla u(z)\right| dz \le |y|\, \|\nabla u\|_{L^1(\Omega)}.$$

Using Corollary 6.1.18, we conclude by density that, for every $u \in S$,

$$\|\tau_y u - u\|_{L^1(\omega)} \le \|\nabla u\|_{L^1(\Omega)}|y| \le \|\nabla u\|_{L^p(\Omega)} m(\Omega)^{1-1/p}|y| \le c_1|y|. \qquad \square$$

**Theorem 6.4.6 (Rellich–Kondrachov Embedding Theorem)** *Let $\Omega$ be a bounded open subset of $\mathbb{R}^N$ of class $C^1$ or the product of $N$ bounded open intervals:*

(a) *If $1 \le p < N$ and $1 \le q < p^*$, then $W^{1,p}(\Omega) \subset L^q(\Omega)$, and the canonical injection is compact.*
(b) *If $N < p < \infty$, then $W^{1,p} \subset C_0(\bar\Omega)$, and the canonical injection is compact.*

**Proof** Let $1 \le p < N, 1 \le q < p^*$. It suffices to use Sobolev's embedding theorem and the preceding lemma.

The case $p > N$ follows from Ascoli's theorem and Sobolev's embedding theorem. $\qquad \square$

We prove three fundamental inequalities.

**Theorem 6.4.7 (Poincaré's Inequality in $W_0^{1,p}$)** *Let $1 \le p < \infty$, and let $\Omega$ be an open subset of $\mathbb{R}^N$ such that $\Omega \subset \mathbb{R}^{N-1} \times ]0, a[$. Then for every $u \in W_0^{1,p}(\Omega)$,*

$$\|u\|_{L^p(\Omega)} \le \frac{a}{2}\|\nabla u\|_{L^p(\Omega)}.$$

**Proof** Let $1 < p < \infty$ and $v \in \mathcal{D}(]0, a[)$. The fundamental theorem of calculus and Hölder's inequality imply that for $0 < x < a$,

$$\left|v(x)\right| \le \frac{1}{2} \int_0^a \left|v'(t)\right| dt \le \frac{a^{1/p'}}{2}\left|\int_0^a \left|v'(t)\right|^p dt\right|^{1/p}.$$

Hence we obtain

$$\int_0^a \left|v(x)\right|^p dx \le \frac{a^{p/p'}}{2^p} a \int_0^a \left|v'(x)\right|^p dx = \frac{a^p}{2^p} \int_0^a \left|v'(x)\right|^p dx.$$

If $u \in \mathcal{D}(\Omega)$, we infer from the preceding inequality and from Fubini's theorem that

$$\int_\Omega |u|^p dx = \int_{\mathbb{R}^{N-1}} dx' \int_0^a |u(x', x_N)|^p dx_N$$

$$\leq \frac{a^p}{2^p} \int_{\mathbb{R}^{N-1}} dx' \int_0^a |\partial_N u(x', x_N)|^p dx_N$$

$$= \frac{a^p}{2^p} \int_\Omega |\partial_N u|^p dx.$$

It is easy to conclude by density. The case $p = 1$ is similar. $\square$

**Definition 6.4.8** A metric space is connected if the only open and closed subsets of $X$ are $\phi$ and $X$.

**Theorem 6.4.9 (Poincaré's Inequality in $W^{1,p}$)** *Let* $1 \leq p < \infty$, *and let* $\Omega$ *be a bounded open connected subset of* $\mathbb{R}^N$. *Assume that* $\Omega$ *is of class* $C^1$. *Then there exists* $c = c(p, \Omega)$, *such that, for every* $u \in W^{1,p}(\Omega)$,

$$\left\| u - \fint u \right\|_{L^p(\Omega)} \leq c \|\nabla u\|_{L^p(\Omega)},$$

*where*

$$\fint u = \frac{1}{m(\Omega)} \int_\Omega u \, dx.$$

*Assume that* $\Omega$ *is convex. Then, for every* $u \in W^{1,p}(\Omega)$,

$$\left\| u - \fint u \right\|_{L^p(\Omega)} \leq 2^{N/p} d \, \|\nabla u\|_{L^p(\Omega)},$$

*where* $d = \sup_{x,y \in \Omega} |x - y|$.

**Proof** Assume that $\Omega$ is of class $C^1$. It suffices to prove that

$$\lambda = \inf \left\{ \|\nabla u\|_p : u \in W^{1,p}(\Omega), \fint u = 0, \|u\|_p = 1 \right\} > 0.$$

Let $(u_n) \subset W^{1,p}(\Omega)$ be a minimizing sequence :

$$\|u_n\|_p = 1, \quad \fint u_n = 0, \quad \|\nabla u_n\|_p \to \lambda.$$

By the Rellich–Kondrachov theorem, we can assume that $u_n \to u$ in $L^p(\Omega)$. Hence $\|u\|_p = 1$ and $\fint u = 0$. If $\lambda = 0$, then, by the closing lemma, $\nabla u = 0$. Since $\Omega$ is connected, $u = \fint u = 0$. This is a contradiction.

Assume now that $\Omega$ is convex and that $u \in C^\infty(\Omega) \cap W^{1,p}(\Omega)$. Hölder's inequality implies that

$$\int_\Omega \left| u(y) - \fint u \right|^p dy \leq \int_\Omega dy \left[ \int_\Omega \frac{|u(x) - u(y)|}{m(\Omega)} dx \right]^p$$

$$\leq \frac{1}{m(\Omega)} \int_\Omega dy \int_\Omega \left| u(x) - u(y) \right|^p dx.$$

It follows from the fundamental theorem of calculus and Hölder's inequality that

$$\int_\Omega dy \int_\Omega \left| u(x) - u(y) \right|^p dx \leq d^p \int_\Omega dy \int_\Omega dx \left[ \int_0^1 \left| \nabla u((1-t)x + ty) \right| dt \right]^p$$

$$\leq d^p \int_\Omega dy \int_\Omega dx \int_0^1 \left| \nabla u((1-t)x + ty) \right|^p dt$$

$$= 2d^p \int_\Omega dy \int_\Omega dx \int_0^{1/2} \left| \nabla u((1-t)x + ty) \right|^p dt$$

$$= 2d^p \int_\Omega dy \int_0^{1/2} dt \int_\Omega \left| \nabla u((1-t)x + ty) \right|^p dx$$

$$\leq 2^N d^p \int_\Omega dy \int_\Omega \left| \nabla u(z) \right|^p dz.$$

We obtain that

$$\int_\Omega \left| u(y) - \fint u \right|^p dy \leq 2^N d^p \int_\Omega \left| \nabla u(y) \right|^p dy.$$

We conclude by density, using Corollary 6.1.18.                                                  □

**Theorem 6.4.10 (Hardy's Inequality)** *Let* $1 < p < N$. *Then for every* $u \in W^{1,p}(\mathbb{R}^N)$, $u/|x| \in L^p(\mathbb{R}^N)$ *and*

$$\|u/|x|\|_{L^p(\mathbb{R}^N)} \leq \frac{p}{N-p} \|\nabla u\|_{L^p(\mathbb{R}^N)}.$$

**Proof** Let $u \in \mathcal{D}(\mathbb{R}^N)$ and $v \in \mathcal{D}(\mathbb{R}^N; \mathbb{R}^N)$. We infer from Lemma 6.1.1 that

$$\int_{\mathbb{R}^N} |u|^p \operatorname{div} v \, dx = -p \int_{\mathbb{R}^N} |u|^{p-2} u \nabla u \cdot v \, dx.$$

Approximating $v(x) = x/|x|^p$ by $v_\varepsilon(x) = x/(|x|^2 + \varepsilon)^{p/2}$, we obtain

$$(N - p) \int_{\mathbb{R}^N} |u|^p / |x|^p dx = -p \int_{\mathbb{R}^N} |u|^{p-2} u \nabla u \cdot x / |x|^p dx.$$

Hölder's inequality implies that

$$\int_{\mathbb{R}^N} |u|^p / |x|^p dx \le \frac{p}{N-p} \left( \int_{\mathbb{R}^N} |u|^{(p-1)p'} / |x|^p dx \right)^{1/p'} \left( \int_{\mathbb{R}^N} |\nabla u|^p dx \right)^{1/p}$$

$$= \frac{p}{N-p} \left( \int_{\mathbb{R}^N} |u|^p / |x|^p dx \right)^{1-1/p} \left( \int_{\mathbb{R}^N} |\nabla u|^p dx \right)^{1/p}.$$

We have thus proved Hardy's inequality in $\mathcal{D}(\mathbb{R}^N)$. Let $u \in W^{1,p}(\mathbb{R}^N)$. Theorem 6.1.10 ensures the existence of a sequence $(u_n) \subset \mathcal{D}(\mathbb{R}^N)$ such that $u_n \to u$ in $W^{1,p}(\mathbb{R}^N)$. We can assume that $u_n \to u$ almost everywhere on $\mathbb{R}^N$. We conclude using Fatou's lemma that

$$||u/|x|||_p \le \lim_{n \to \infty} ||u_n/|x|||_p \le \frac{p}{N-p} \lim_{n \to \infty} ||\nabla u_n||_p = \frac{p}{N-p} ||\nabla u||_p. \quad \square$$

Fractional Sobolev spaces are interpolation spaces between $L^p(\Omega)$ and $W^{1,p}(\Omega)$.

**Definition 6.4.11** Let $1 \le p < \infty$, $0 < s < 1$, and $u \in L^p(\Omega)$. We define

$$|u|_{W^{s,p}(\Omega)} = |u|_{s,p} = \left( \int_\Omega \int_\Omega \frac{|u(x) - u(y)|^p}{|x-y|^{N+sp}} dx dy \right)^{1/p} \le +\infty.$$

On the fractional Sobolev space

$$W^{s,p}(\Omega) = \{ u \in L^p(\Omega) : |u|_{W^{s,p}(\Omega)} < +\infty \},$$

we define the norm

$$||u||_{W^{s,p}(\Omega)} = ||u||_{s,p} = ||u||_{L^p(\Omega)} + |u|_{W^{s,p}(\Omega)}.$$

We give, without proof, the characterization of traces due to Gagliardo [26].

**Theorem 6.4.12** Let $1 < p < \infty$.

(a) For every $u \in W^{1,p}(\mathbb{R}^N)$, $\gamma_0 u \in W^{1-1/p,p}(\mathbb{R}^{N-1})$.
(b) The mapping $\gamma_0 : W^{1,p}(\mathbb{R}^N) \to W^{1-1/p,p}(\mathbb{R}^{N-1})$ is continuous and surjective.
(c) The mapping $\gamma_0 : W^{1,1}(\mathbb{R}^N) \to L^1(\mathbb{R}^{N-1})$ is continuous and surjective.

## 6.5  Comments

The main references on Sobolev spaces are the books:

– R. Adams and J. Fournier, *Sobolev spaces* [1]
– H. Brezis, *Analyse fonctionnelle, théorie et applications* [8]
– V. Maz'ya, *Sobolev spaces with applications to elliptic partial differential equations* [51]

Our proof of the *trace inequality* follows closely:

– A.C. Ponce, *Elliptic PDEs, measures, and capacities*, European Mathematical Society, 2016

The theory of partial differential equations was at the origin of Sobolev spaces. We recommend [9] on the history of partial differential equations and [55] on the prehistory of Sobolev spaces.

Because of Poincaré's inequalities, for every smooth, bounded open connected set $\Omega$, we have that

$$\lambda_1(\Omega) = \inf\left\{ \int_\Omega |\nabla u|^2 dx : u \in H_0^1(\Omega), \int_\Omega u^2 dx = 1 \right\} > 0,$$

$$\mu_2(\Omega) = \inf\left\{ \int_\Omega |\nabla u|^2 dx : u \in H^1(\Omega), \int_\Omega u^2 dx = 1, \int_\Omega u\, dx = 0 \right\} > 0.$$

Hence the first eigenvalue $\lambda_1(\Omega)$ of Dirichlet's problem

$$\begin{cases} -\Delta u = \lambda u & \text{in } \Omega, \\ u = 0 & \text{on } \partial\Omega, \end{cases}$$

and the second eigenvalue $\mu_2(\Omega)$ of the Neumann problem

$$\begin{cases} -\Delta u = \lambda u & \text{in } \Omega, \\ n \cdot \nabla u = 0 & \text{on } \partial\Omega, \end{cases}$$

are strictly positive. Let us denote by $B$ an open ball such that $m(B) = m(\Omega)$. Then

$$\lambda_1(B) \le \lambda_1(\Omega) \quad \text{(Faber–Krahn inequality)},$$
$$\mu_2(\Omega) \le \mu_2(B) \quad \text{(Weinberger, 1956)}.$$

Moreover, if $\Omega$ is convex with diameter $d$, then

$$\pi^2/d^2 \le \mu_2(\Omega) \quad \text{(Payne–Weinberger, 1960)}.$$

We prove in Theorem 6.4.9 the weaker estimate

$$1/(2^N d^2) \leq \mu_2(\Omega).$$

There exists a bounded, connected open set $\Omega \subset \mathbb{R}^2$ such that $\mu_2(\Omega) = 0$. Consider on two sides of a square $Q$, two infinite sequences of small squares connected to $Q$ by very thin pipes.

## 6.6 Exercises for Chap. 6

1. Let $\Omega = B(0, 1) \subset \mathbb{R}^N$. Then for $\lambda \neq 0$,

$$(\lambda - 1)p + N > 0 \iff |x|^\lambda \in W^{1,p}(\Omega),$$

$$\lambda p + N < 0 \iff |x|^\lambda \in W^{1,p}(\mathbb{R}^N \setminus \overline{\Omega}),$$

$$p < N \iff \frac{x}{|x|} \in W^{1,p}(\Omega; \mathbb{R}^N).$$

2. Let $1 < p < \infty$ and $u \in L^p(\Omega)$. The following properties are equivalent:
   (a) $u \in W^{1,p}(\Omega)$;
   (b) $\sup \left\{ \int_\Omega u \operatorname{div} v \, dx : v \in \mathcal{D}(\Omega, \mathbb{R}^N), \|v\|_{L^{p'}(\Omega)} = 1 \right\} < \infty$;
   (c) there exists $c > 0$ such that for every $\omega \subset\subset \Omega$ and for every $y \in \mathbb{R}^N$ such that $|y| < d(\omega, \partial\Omega)$,

$$\|\tau_y u - u\|_{L^p(\omega)} \leq c|y|.$$

3. Let $1 \leq p < N$ and let $\Omega$ be an open subset of $\mathbb{R}^N$. Define

$$S(\Omega) = \inf_{\substack{u \in \mathcal{D}(\Omega) \\ \|u\|_{L^{p^*}(\Omega)} = 1}} \|\nabla u\|_{L^p(\Omega)}.$$

Then $S(\Omega) = S(\mathbb{R}^N)$.

4. Let $1 \leq p < N$. Then

$$\frac{1}{2^N} S(\mathbb{R}^N) = \inf \left\{ \|\nabla u\|_{L^p(\mathbb{R}^N_+)} / \|u\|_{L^{p^*}(\mathbb{R}^N_+)} : u \in H^1(\mathbb{R}^N_+) \setminus \{0\} \right\}.$$

5. Poincaré–Sobolev inequality.

   (a) Let $1 < p < N$, and let $\Omega$ be an open bounded connected subset of $\mathbb{R}^N$ of class $C^1$. Then there exists $c > 0$ such that for every $u \in W^{1,p}(\Omega)$,

$$\left\|u - \fint u\right\|_{L^{p^*}(\Omega)} \le c\|\nabla u\|_{L^p(\Omega)},$$

where $\fint u = \dfrac{1}{m(\Omega)}\displaystyle\int_\Omega u\,dx$. *Hint*: Apply Theorem 6.4.4 to $u - \fint u$.

(b)  Let $A = \{u = 0\}$ and assume that $m(A) > 0$. Then

$$\|u\|_{L^{p^*}(\Omega)} \le c\left(1 + \left[\frac{m(\Omega)}{m(A)}\right]^{1/p^*}\right)\|\nabla u\|_{L^p(\Omega)}.$$

*Hint*:

$$\left|\fint u\right| m(A)^{1/p^*} \le \left\|u - \fint u\right\|_{L^{p^*}(\Omega)}.$$

6.  Nash's inequality. Let $N \ge 3$. Then for every $u \in \mathcal{D}(\mathbb{R}^N)$,

$$\|u\|_2^{2+4/N} \le c\|u\|_1^{4/N}\|\nabla u\|_2^2.$$

*Hint*: Use the interpolation inequality.

7.  Let $1 \le p < N$ and $q = p(N-1)/(N-p)$. Then for every $u \in \mathcal{D}(\overline{\mathbb{R}^N_+})$,

$$\int_{\mathbb{R}^{N-1}} |u(x',0)|^q dx' \le q\|u\|_{L^{p^*}(\mathbb{R}^N_+)}^{q-1}\|\partial_N u\|_{L^p(\mathbb{R}^N_+)}.$$

8.  Verify that Hardy's inequality is optimal using the family

$$\begin{aligned} u_\varepsilon(x) &= 1, & |x| \le 1, \\ &= |x|^{\frac{p-N}{p}-\varepsilon}, & |x| > 1. \end{aligned}$$

9.  Let $1 \le p < N$. Then $\mathcal{D}(\mathbb{R}^N \setminus \{0\})$ is dense in $W^{1,p}(\mathbb{R}^N)$.

10.  Let $2 \le N < p < \infty$. Then for every $u \in W_0^{1,p}(\mathbb{R}^N \setminus \{0\})$, $u/|x| \in L^p(\mathbb{R}^N)$ and

$$\|u/|x|\|_{L^p(\mathbb{R}^N)} \le \frac{p}{p-N}\|\nabla u\|_{L^p(\mathbb{R}^N)}.$$

11.  Let $1 \le p < \infty$. Verify that the embedding $W^{1,p}(\mathbb{R}^N) \subset L^p(\mathbb{R}^N)$ is not compact. Let $1 \le p < N$. Verify that the embedding $W_0^{1,p}(B(0,1)) \subset L^{p^*}(B(0,1))$ is not compact.

12.  Let us denote by $\mathcal{D}_r(\mathbb{R}^N)$ the space of radial functions in $\mathcal{D}(\mathbb{R}^N)$. Let $N \ge 2$ and $1 \le p < \infty$. Then there exists $c(N,p) > 0$ such that for every $u \in \mathcal{D}_r(\mathbb{R}^N)$,

$$\left|u(x)\right| \le c(N, p)\|u\|_p^{1/p'}\|\nabla u\|_p^{1/p}|x|^{(1-N)/p}.$$

Let $1 \le p < N$. Then there exists $d(N, p) > 0$ such that for every $u \in \mathcal{D}_r(\mathbb{R}^N)$,

$$\left|u(x)\right| \le d(N, p)\|\nabla u\|_p|x|^{(p-N)/p}.$$

*Hint*: Let us write $u(x) = u(r)$, $r = |x|$, so that

$$r^{N-1}\left|u(r)\right|^p \le p\int_r^\infty \left|u(s)\right|^{p-1}\left|\frac{du}{dr}(s)\right|s^{N-1}ds,$$

$$\left|u(r)\right| \le \int_r^\infty \left|\frac{du}{dr}(s)\right|ds.$$

13. Let us denote by $W_r^{1,p}(\mathbb{R}^N)$ the space of radial functions in $W^{1,p}(\mathbb{R}^N)$. Verify that the space $\mathcal{D}_r(\mathbb{R}^N)$ is dense in $W_r^{1,p}(\mathbb{R}^N)$.

14. Let $1 \le p < N$ and $p < q < p^*$. Verify that the embedding $W_r^{1,p}(\mathbb{R}^N) \subset L^q(\mathbb{R}^N)$ is compact. Verify also that the embedding $W_r^{1,p}(\mathbb{R}^N) \subset L^p(\mathbb{R}^N)$ is not compact.

15. Let $1 \le p < \infty$ and let $\Omega$ be an open subset of $\mathbb{R}^N$. Prove that the map

$$W^{1,p}(\Omega) \to W^{1,p}(\Omega) : u \mapsto u^+$$

is continuous. *Hint*: $\nabla u^+ = H(u)\nabla u$, where

$$H(t) = 1, \quad t > 0,$$
$$= 0, \quad t \le 0.$$

16. Sobolev implies Poincaré. Let $\Omega$ be an open subset of $\mathbb{R}^N$ ($N \ge 2$) such that $m(\Omega) < +\infty$, and let $1 \le p < +\infty$. Then there exists $c = c(p, N)$ such that, for every $u \in W_0^{1,p}(\Omega)$,

$$\|u\|_p \le c\, m(\Omega)^{1/N}\|\nabla u\|_p.$$

*Hint*. (a) If $1 \le p < N$, then

$$\|u\|_p \le m(\Omega)^{1/N}\|u\|_{p^*} \le c\, m(\Omega)^{1/N}\|\nabla u\|_p.$$

(b) If $p \ge N$, then

$$\|u\|_p = \|u\|_{q^*} \le c\|\nabla u\|_q \le c\, m(\Omega)^{1/N}\|\nabla u\|_p.$$

17. Let $\Omega$ be an open bounded convex subset of $\mathbb{R}^N$, $N \geq 2$, and $u \in C^1(\Omega) \cap W^{1,1}(\Omega)$. Then, for every $x \in \Omega$,

$$\left| u(x) - \fint u \right| \leq \frac{1}{N} \frac{d^N}{m(\Omega)} \int_\Omega \frac{|\nabla u(y)|}{|y-x|^{N-1}} dy,$$

where $\fint u = \dfrac{1}{m(\Omega)} \displaystyle\int_\Omega u(x)dx$ and $d = \sup\limits_{x,y \in \Omega} |y-x|$.

*Hint.* Define

$$v(y) = |\nabla u(y)| \quad , y \in \Omega,$$
$$= 0 \qquad\qquad , y \in \mathbb{R}^N \setminus \Omega.$$

(a) $u(x) - u(y) = \displaystyle\int_0^{|y-x|} \nabla u(x + r\sigma) \cdot \sigma\, dr$, $\sigma = \dfrac{y-x}{|y-x|}$.

(b)

$$m(\Omega)\left| u(x) - \fint u \right| \leq \int_\Omega dy \int_0^{|y-x|} v(x + r\sigma)dr$$

$$= \int_{\omega - x} dz \int_0^{|z|} v\left( x + r\frac{z}{|z|} \right) dr$$

$$\leq \int_{\mathbb{S}^{N-1}} d\sigma \int_0^d \rho^{N-1}d\rho \int_0^\infty v(x + r\sigma)dr$$

$$= \frac{d^N}{N} \int_{\mathbb{R}^N} \frac{v(x + z)}{|z|^{N-1}} dz.$$

18. Let us define, for every bounded connected open subset $\Omega$ of $\mathbb{R}^N$, and for $1 \leq p < \infty$,

$$\lambda(p, \Omega) = \inf\left\{ \|\nabla u\|_p : u \in W^{1,p}(\Omega),\ \fint u = 0,\ \|u\|_p = 1 \right\}.$$

For every $1 \leq p < \infty$, there exists a bounded connected open subset $\Omega$ of $\mathbb{R}^2$ such that $\lambda(p, \Omega) = 0$.
*Hint.* Consider on two sides of a square $Q$ two infinite sequences of small squares connected to $Q$ by very thin pipes.

19. Prove that, for every $1 \leq p < \infty$,

$$\inf\left\{ \lambda(p, \Omega): \Omega \text{ is a smooth bounded connected open subset of } \mathbb{R}^2, m(\Omega) = 1 \right\} = 0.$$

*Hint.* Consider a sequence of pairs of disks smoothly connected by very thin pipes.

20. Generalized Poincaré's inequality. Let $1 \leq p < \infty$, let $\Omega$ be a smooth bounded connected open subset of $\mathbb{R}^N$, and let $f \in [W^{1,p}(\Omega)]^*$ be such that

$$< f, 1 > = 1.$$

Then there exists $c > 0$ such that, for every $u \in W^{1,p}(\Omega)$,

$$\|u - < f, u > \|_p \leq c \|\nabla u\|_p.$$

# Chapter 7
# Capacity

## 7.1 Capacity

The notion of *capacity* appears in potential theory. The abstract theory was formulated by Choquet in 1954. In this section, we denote by $X$ a metric space, by $\mathcal{K}$ the class of compact subsets of $X$, and by $O$ the class of open subsets of $X$.

**Definition 7.1.1** A capacity on $X$ is a function

$$\text{cap} : \mathcal{K} \to [0, +\infty] : K \to \text{cap}(K)$$

such that:

$(C_1)$ (monotonicity.) For every $A, B \in \mathcal{K}$ such that $A \subset B$, $\text{cap}(A) \leq \text{cap}(B)$.
$(C_2)$ (regularity.) For every $K \in \mathcal{K}$ and for every $a > \text{cap}(K)$, there exists $U \in O$ such that $K \subset U$, and for all $C \in \mathcal{K}$ satisfying $C \subset U$, $\text{cap}(C) < a$.
$(C_3)$ (strong subadditivity.) For every $A, B \in \mathcal{K}$,

$$\text{cap}(A \cup B) + \text{cap}(A \cap B) \leq \text{cap}(A) + \text{cap}(B).$$

The Lebesgue measure of a compact subset of $\mathbb{R}^N$ is a capacity.
We denote by cap a capacity on $X$. We extend the capacity to the open subsets of $X$.

**Definition 7.1.2** The capacity of $U \in O$ is defined by

$$\text{cap}(U) = \sup\{\text{cap}(K) : K \in \mathcal{K} \text{ and } K \subset U\}.$$

**Lemma 7.1.3** *Let $A, B \in O$ and $K \in \mathcal{K}$ be such that $K \subset A \cup B$. Then there exist $L, M \in \mathcal{K}$ such that $L \subset A$, $M \subset B$, and $K = L \cup M$.*

© Springer Nature Switzerland AG 2022
M. Willem, *Functional Analysis*, Cornerstones,
https://doi.org/10.1007/978-3-031-09149-0_7

***Proof*** The compact sets $K \setminus A$ and $K \setminus B$ are disjoint. Hence there exist disjoint open sets $U$ and $V$ such that $K \setminus A \subset U$ and $K \setminus B \subset V$. It suffices to define $L = K \setminus U$ and $M = K \setminus V$.                                      □

**Proposition 7.1.4**

*(a) (monotonicity.) For every $A, B \in O$ such that $A \subset B$, $\text{cap}(A) \leq \text{cap}(B)$.*
*(b) (regularity.) For every $K \in \mathcal{K}$, $\text{cap}(K) = \inf\{\text{cap}(U) : U \in O \text{ and } U \supset K\}$.*
*(c) (strong subadditivity.) For every $A, B \in O$,*

$$\text{cap}(A \cup B) + \text{cap } A \cap B) \leq \text{cap}(A) + \text{cap}(B).$$

*Proof*

(a) Monotonicity is clear.
(b) Let us define $\text{Cap}(K) = \inf\{\text{cap}(U) : U \in O \text{ and } U \supset K\}$. By definition, $\text{cap}(K) \leq \text{Cap}(K)$. Let $a > \text{cap}(K)$. There exists $U \in O$ such that $K \subset U$ and for every $C \in \mathcal{K}$ satisfying $C \subset U$, $\text{cap}(C) < a$. Hence $\text{Cap}(K) \leq \text{cap}(U) < a$. Since $a > \text{cap}(K)$ is arbitrary, we conclude that $\text{Cap}(K) \leq \text{cap}(K)$.
(c) Let $A, B \in O$, $a < \text{cap}(A \cup B)$, and $b < \text{cap}(A \cap B)$. By definition, there exist $K, C \in \mathcal{K}$ such that $K \subset A \cup B$, $C \subset A \cap B$, $a < \text{cap}(K)$, and $b \leq \text{cap}(C)$. We can assume that $C \subset K$. The preceding lemma implies the existence of $L, M \in \mathcal{K}$ such that $L \subset A$, $M \subset B$, and $K = L \cup M$. We can assume that $C \subset L \cap M$. We obtain by monotonicity and strong subadditivity that

$$a + b \leq \text{cap}(K) + \text{cap}(C) \leq \text{cap}(L \cup M) + \text{cap}(L \cap M)$$
$$\leq \text{cap}(L) + \text{cap}(M) \leq \text{cap}(A) + \text{cap}(B).$$

Since $a < \text{cap}(A \cup B)$ and $b < \text{cap}(A \cap B)$ are arbitrary, the proof is complete.   □

We extend the capacity to all subsets of $X$.

**Definition 7.1.5** The capacity of a subset $A$ of $X$ is defined by

$$\text{cap}(A) = \inf\{\text{cap}(U) : U \in O \text{ and } U \supset A\}.$$

By regularity, the capacity of compact subsets is well defined.

**Proposition 7.1.6**

*(a) (monotonicity). For every $A, B \subset X$, $\text{cap}(A) \leq \text{cap}(B)$.*
*(b) (strong subadditivity). For every $A, B \subset X$,*

$$\text{cap}(A \cup B) + \text{cap}(A \cap B) \leq \text{cap}(A) + \text{cap}(B).$$

## Proof

(a) Monotonicity is clear.

(b) Let $A, B \subset X$ and $U, V \in O$ be such that $A \subset U$ and $B \subset V$. We have

$$\operatorname{cap}(A \cup B) + \operatorname{cap}(A \cap B) \leq \operatorname{cap}(U \cup V) + \operatorname{cap}(U \cap V) \leq \operatorname{cap}(U) + \operatorname{cap}(V).$$

It is easy to conclude the proof. □

**Proposition 7.1.7** *Let* $(K_n)$ *be a decreasing sequence in* $\mathcal{K}$. *Then*

$$\operatorname{cap}\left(\bigcap_{n=1}^{\infty} K_n\right) = \lim_{n \to \infty} \operatorname{cap}(K_n).$$

**Proof** Let $K = \bigcap_{n=1}^{\infty} K_n$ and $U \in O$, $U \supset K$. By compactness, there exists $m$ such that $K_m \subset U$. We obtain, by monotonicity, $\operatorname{cap}(K) \leq \lim_{n \to \infty} \operatorname{cap}(K_n) \leq \operatorname{cap}(U)$. It suffices then to take the infimum with respect to $U$. □

**Lemma 7.1.8** *Let* $(U_n)$ *be an increasing sequence in* $O$. *Then*

$$\operatorname{cap}\left(\bigcup_{n=1}^{\infty} U_n\right) = \lim_{n \to \infty} \operatorname{cap}(U_n).$$

**Proof** Let $U = \bigcup_{n=1}^{\infty} U_n$ and $K \in \mathcal{K}$, $K \subset U$. By compactness, there exists $m$ such that $K \subset U_m$. We obtain by monotonicity $\operatorname{cap}(K) \leq \lim_{n \to \infty} \operatorname{cap}(U_n) \leq \operatorname{cap}U$. It suffices then to take the supremum with respect to $K$. □

**Theorem 7.1.9** *Let* $(A_n)$ *be an increasing sequence of subsets of* $X$. *Then*

$$\operatorname{cap}\left(\bigcup_{n=1}^{\infty} A_n\right) = \lim_{n \to \infty} \operatorname{cap}(A_n).$$

**Proof** Let $A = \bigcup_{n=1}^{\infty} A_n$. By monotonicity, $\lim_{n \to \infty} \operatorname{cap}(A_n) \leq \operatorname{cap}(A)$. We can assume that $\lim_{n \to \infty} \operatorname{cap}(A_n) < +\infty$. Let $\varepsilon > 0$ and $a_n = 1 - 1/(n + 1)$. We construct, by induction, an increasing sequence $(U_n) \subset O$ such that $A_n \subset U_n$ and

$$\operatorname{cap}(U_n) \leq \operatorname{cap}(A_n) + \varepsilon\, a_n. \tag{$*$}$$

When $n = 1$, $(*)$ holds by definition. Assume that $(*)$ holds for $n$. By definition, there exists $V \in O$ such that $A_{n+1} \subset V$ and

$$\text{cap}(V) \leq \text{cap}(A_{n+1}) + \varepsilon(a_{n+1} - a_n).$$

We define $U_{n+1} = U_n \cup V$, so that $A_{n+1} \subset U_{n+1}$. We obtain, by strong subadditivity,

$$\begin{aligned}
\text{cap}(U_{n+1}) &\leq \text{cap}(U_n) + \text{cap}(V) - \text{cap}(U_n \cap V) \\
&\leq \text{cap}(A_n) + \varepsilon\, a_n + \text{cap}(A_{n+1}) + \varepsilon(a_{n+1} - a_n) - \text{cap}(A_n) \\
&= \text{cap}(A_{n+1}) + \varepsilon\, a_{n+1}.
\end{aligned}$$

It follows from $(*)$ and the preceding lemma that

$$\text{cap}(A) \leq \text{cap}\left(\bigcup_{n=1}^{\infty} U_n\right) = \lim_{n\to\infty} \text{cap}(U_n) \leq \lim_{n\to\infty} \text{cap}(A_n) + \varepsilon.$$

Since $\varepsilon > 0$ is arbitrary, the proof is complete.                                    $\square$

**Corollary 7.1.10 (Countable Subadditivity)** *Let* $(A_n)$ *be a sequence of subsets of* $X$. *Then* $\text{cap}\left(\bigcup_{n=1}^{\infty} A_n\right) \leq \sum_{n=1}^{\infty} \text{cap}(A_n).$

**Proof** Let $B_k = \bigcup_{n=1}^{k} A_k$. We have

$$\text{cap}\left(\bigcup_{n=1}^{\infty} A_n\right) = \text{cap}\left(\bigcup_{k=1}^{\infty} B_k\right) = \lim_{k\to\infty} \text{cap}(B_k) \leq \sum_{n=1}^{\infty} \text{cap}(A_n).    \qquad \square$$

**Definition 7.1.11** The outer Lebesgue measure of a subset of $\mathbb{R}^N$ is defined by

$$m^*(A) = \inf\{m(U) : U \text{ is open and } U \supset A\}.$$

## 7.2   Variational Capacity

In order to define *variational capacity*, we introduce the space $\mathcal{D}^{1,p}(\mathbb{R}^N)$.

**Definition 7.2.1** Let $1 \leq p < N$. On the space

$$\mathcal{D}^{1,p}(\mathbb{R}^N) = \{u \in L^{p^*}(\mathbb{R}^N) : \nabla u \in L^p(\mathbb{R}^N; \mathbb{R}^N)\},$$

we define the norm

$$||u||_{\mathcal{D}^{1,p}(\mathbb{R}^N)} = ||\nabla u||_p.$$

**Proposition 7.2.2** *Let* $1 \leq p < N$.

(a) *The space* $\mathcal{D}(\mathbb{R}^N)$ *is dense in* $\mathcal{D}^{1,p}(\mathbb{R}^N)$.
(b) *(Sobolev's inequality.) There exists* $c = c(p, N)$ *such that for every* $u \in \mathcal{D}^{1,p}(\mathbb{R}^N)$,

$$||u||_{p^*} \leq c||\nabla u||_p.$$

(c) *The space* $\mathcal{D}^{1,p}(\mathbb{R}^N)$ *is complete.*

*Proof* The space $\mathcal{D}(\mathbb{R}^N)$ is dense in $\mathcal{D}^{1,p}(\mathbb{R}^N)$ with the norm $||u||_{p^*} + ||\nabla u||_p$. The argument is similar to that of the proof of Theorem 6.1.10.

Sobolev's inequality follows by density from Lemma 6.4.2. Hence for every $u \in \mathcal{D}^{1,p}(\mathbb{R}^N)$,

$$||\nabla u||_p \leq ||u||_{p^*} + ||\nabla u||_p \leq (c+1)||\nabla u||_p.$$

Let $(u_n)$ be a Cauchy sequence in $\mathcal{D}^{1,p}(\mathbb{R}^N)$. Then $u_n \to u$ in $L^{p^*}(\mathbb{R}^N)$, and for $1 \leq k \leq N$, $\partial_k u_n \to v_k$ in $L^p(\mathbb{R}^N)$. By the closing lemma, for $1 \leq k \leq N$, $\partial_k u = v_k$. We conclude that $u_n \to u$ in $\mathcal{D}^{1,p}(\mathbb{R}^N)$. □

**Proposition 7.2.3** *Every bounded sequence in* $\mathcal{D}^{1,p}(\mathbb{R}^N)$ *contains a subsequence converging in* $L^1_{loc}(\mathbb{R}^N)$ *and almost everywhere on* $\mathbb{R}^N$.

*Proof* Cantor's diagonal argument will be used. Let $(u_n)$ be bounded in $\mathcal{D}^{1,p}(\mathbb{R}^N)$. By Sobolev's inequality, for every $k \geq 1$, $(u_n)$ is bounded in $W^{1,1}(B(0, k))$. Rellich's theorem and Proposition 4.2.10 imply the existence of a subsequence $(u_{1,n})$ of $(u_n)$ converging in $L^1(B(0, 1))$ and almost everywhere on $B(0, 1)$. By induction, for every $k$, there exists a subsequence $(u_{k,n})$ of $(u_{k-1,n})$ converging in $L^1(B(0, k))$ and almost everywhere on $B(0, k)$. The sequence $v_n = u_{n,n}$ converges in $L^1_{loc}(\mathbb{R}^N)$ and almost everywhere on $\mathbb{R}^N$. □

**Definition 7.2.4** Let $1 \leq p < N$ and let $K$ be a compact subset of $\mathbb{R}^N$. The capacity of degree $p$ of $K$ is defined by

$$\text{cap}_p(K) = \inf \left\{ \int_{\mathbb{R}^N} |\nabla u|^p dx : u \in \mathcal{D}_K^{1,p}(\mathbb{R}^N) \right\},$$

where

$$\mathcal{D}_K^{1,p}(\mathbb{R}^N) = \{u \in \mathcal{D}^{1,p}(\mathbb{R}^N) : \text{there exists } U \text{ open such that } K \subset U \text{ and } \chi_U \leq u$$

almost everywhere}.

**Theorem 7.2.5** *The capacity of degree p is a capacity on* $\mathbb{R}^N$.

*Proof*

(a)  Monotonicity is clear by definition.
(b)  Let $K$ be compact and $a > \mathrm{cap}_p(\mathbb{R}^N)$. There exist $u \in \mathcal{D}^{1,p}(\mathbb{R}^N)$ and $U$ open such that $K \subset U$, $\chi_U \le u$ almost everywhere, and $\int_{\mathbb{R}^N} |\nabla u|^p dx < a$. For every compact set $C \subset U$, we have

$$\mathrm{cap}_p(C) \le \int_{\mathbb{R}^N} |\nabla u|^p dx < a,$$

so that $\mathrm{cap}_p$ is regular.

(c)  Let $A$ and $B$ be compact sets, $a > \mathrm{cap}_p(A)$, and $b > \mathrm{cap}_p(B)$. There exist $u, v \in \mathcal{D}^{1,p}(\mathbb{R}^N)$ and $U$ and $V$ open sets such that $A \subset U$, $B \subset V$, $\chi_U \le u$, and $\chi_V \le v$ almost everywhere and

$$\int_{\mathbb{R}^N} |\nabla u|^p dx < a, \qquad \int_{\mathbb{R}^N} |\nabla v|^p dx < b.$$

Since $\max(u, v) \in \mathcal{D}^{1,p}_{A \cup B}(\mathbb{R}^N)$ and $\min(u, v) \in \mathcal{D}^{1,p}_{A \cap B}(\mathbb{R}^N)$, Corollary 6.1.14 implies that

$$\int_{\mathbb{R}^N} |\nabla \max(u, v)|^p dx + \int_{\mathbb{R}^N} |\nabla \min(u, v)|^p = \int_{\mathbb{R}^N} |\nabla u|^p dx + \int_{\mathbb{R}^N} |\nabla v|^p dx \le a + b.$$

We conclude that

$$\mathrm{cap}_p(A \cup B) + \mathrm{cap}_p(A \cap B) \le a + b.$$

Since $a > \mathrm{cap}_p(A)$ and $b > \mathrm{cap}_p(B)$ are arbitrary, $\mathrm{cap}_p$ is strongly subadditive.

□

The variational capacity is finer than the Lebesgue measure.

**Proposition 7.2.6** *There exists a constant $c = c(p, N)$ such that for every $A \subset \mathbb{R}^N$,*

$$m^*(A) \le c \, \mathrm{cap}_p(A)^{N/(N-p)}.$$

*Proof* Let $K$ be a compact set and $u \in \mathcal{D}^{1,p}_K(\mathbb{R}^N)$. It follows from Sobolev's inequality that

$$m(K) \le \int_{\mathbb{R}^N} |u|^{p^*} dx \le c \left( \int_{\mathbb{R}^N} |\nabla u|^p dx \right)^{p^*/p}.$$

By definition,

$$m(K) \leq c \, \mathrm{cap}_p(K)^{N/(N-p)}.$$

To conclude, it suffices to extend this inequality to open subsets of $\mathbb{R}^N$ and to arbitrary subsets of $\mathbb{R}^N$. □

The variational capacity differs essentially from the Lebesgue measure.

**Proposition 7.2.7** *Let $K$ be a compact set. Then*

$$\mathrm{cap}_p(\partial K) = \mathrm{cap}_p(K).$$

*Proof* Let $a > \mathrm{cap}_p(\partial K)$. There exist $u \in \mathcal{D}^{1,p}(\mathbb{R}^N)$ and an open set $U$ such that $\partial K \subset U$, $\chi_U \leq u$ almost everywhere, and

$$\int_{\mathbb{R}^N} |\nabla u|^p dx < a.$$

Let us define $V = U \cup K$ and $v = \max(u, \chi_V)$. Then $v \in \mathcal{D}_K^{1,p}(\mathbb{R}^N)$ and

$$\int_{\mathbb{R}^N} |\nabla v|^p dx \leq \int_{\mathbb{R}^N} |\nabla u|^p dx,$$

so that $\mathrm{cap}_p(K) < a$. Since $a > \mathrm{cap}_p(\partial K)$ is arbitrary, we obtain

$$\mathrm{cap}_p(K) \leq \mathrm{cap}_p(\partial K) \leq \mathrm{cap}_p(K). \qquad □$$

*Example* Let $1 \leq p < N$ and let $B$ be a closed ball in $\mathbb{R}^N$. We deduce from the preceding propositions that

$$0 < \mathrm{cap}_p(B) = \mathrm{cap}_p(\partial B).$$

**Theorem 7.2.8** *Let $1 < p < N$ and $U$ an open set. Then*

$$\mathrm{cap}_p(U) = \inf \left\{ \int_{\mathbb{R}^N} |\nabla u|^p dx : u \in \mathcal{D}^{1,p}(\mathbb{R}^N), \, \chi_U \leq u \text{ almost everywhere} \right\}.$$

*Proof* Let us denote by $\mathrm{Cap}_p(U)$ the second member of the preceding equality. It is clear by definition that $\mathrm{cap}_p(U) \leq \mathrm{Cap}_p(U)$.

Assume that $\mathrm{cap}_p(U) < \infty$. Let $(K_n)$ be an increasing sequence of compact subsets of $U$ such that $U = \bigcup_{n=1}^{\infty} K_n$, and let $(u_n) \subset \mathcal{D}^{1,p}(\mathbb{R}^N)$ be such that for every $n$, $\chi_{K_n} \le u_n$ almost everywhere and

$$\int_{\mathbb{R}^N} |\nabla u_n|^p dx \le \mathrm{cap}_p(K_n) + 1/n.$$

The sequence $(u_n)$ is bounded in $\mathcal{D}^{1,p}(\mathbb{R}^N)$. By Proposition 7.2.3, we can assume that $u_n \to u$ in $L^1_{\mathrm{loc}}(\mathbb{R}^N)$ and almost everywhere. It follows from Sobolev's inequality that $u \in L^{p^*}(\mathbb{R}^N)$. Theorem 6.1.7 implies that

$$\int_{\mathbb{R}^N} |\nabla u|^p dx \le \varliminf_{n \to \infty} \int_{\mathbb{R}^N} |\nabla u_n|^p dx \le \lim_{n \to \infty} \mathrm{cap}_p(K_n) \le \mathrm{cap}_p(U).$$

By Theorem 7.1.9, $\lim_{n \to \infty} \mathrm{cap}_p(K_n) = \mathrm{cap}_p(U)$.) Since almost everywhere, $\chi_U \le u$, we conclude that $\mathrm{Cap}_p(U) \le \mathrm{cap}_p(U)$.                                      $\square$

**Corollary 7.2.9** *Let* $1 < p < N$, *and let* $U$ *and* $V$ *be open sets such that* $U \subset V$ *and* $m(V \setminus U) = 0$. *Then* $\mathrm{cap}_p(U) = \mathrm{cap}_p(V)$.

**Proof** Let $u \in \mathcal{D}^{1,p}(\mathbb{R}^N)$ be such that $\chi_U \le u$ almost everywhere. Then $\chi_V \le u$ almost everywhere.                                      $\square$

**Corollary 7.2.10 (Capacity Inequality)** *Let* $1 < p < N$ *and* $u \in \mathcal{D}(\mathbb{R}^N)$. *Then for every* $t > 0$,

$$\mathrm{cap}_p(\{|u| > t\}) \le t^{-p} \int_{\mathbb{R}^N} |\nabla u|^p dx.$$

**Proof** By Corollary 6.1.14, $|u|/t \in \mathcal{D}^{1,p}(\mathbb{R}^N)$.                                      $\square$

**Definition 7.2.11** Let $1 \le p < N$. A function $v : \mathbb{R}^N \to \mathbb{R}$ is quasicontinuous of degree $p$ if for every $\varepsilon > 0$, there exists an open set such that $\mathrm{cap}_p(\omega) \le \varepsilon$ and $v|_{\mathbb{R}^N \setminus \omega}$ is continuous. Two quasicontinuous functions of degree $p$, $v$, and $w$ are equal quasi-everywhere if $\mathrm{cap}_p(\{|v - w| > 0\}) = 0$.

**Proposition 7.2.12** *Let* $1 < p < N$ *and let* $v$ *and* $w$ *be quasicontinuous functions of degree* $p$ *and almost everywhere equal. Then* $v$ *and* $w$ *are quasi-everywhere equal.*

**Proof** By assumption, $m(A) = 0$, where $A = \{|v - w| > 0\}$, and for every $n$, there exists an open set such that $\mathrm{cap}_p(\omega_n) \le 1/n$ and $|v - w||_{\mathbb{R}^N \setminus \omega_n}$ are continuous. It follows that $A \cup \omega_n$ is open. We conclude, using Corollary 7.2.9, that

□

$$\text{cap}_p(A) \le \text{cap}_p(A \cup \omega_n) = \text{cap}_p(\omega_n) \to 0, \quad n \to \infty.$$

**Proposition 7.2.13** *Let* $1 < p < N$ *and* $u \in \mathcal{D}^{1,p}(\mathbb{R}^N)$. *Then there exists a function* $v$ *quasicontinuous of degree* $p$ *and almost everywhere equal to* $u$.

**Proof** By Proposition 7.2.2, there exists $(u_n) \subset \mathcal{D}(\mathbb{R}^N)$ such that $u_n \to u$ in $\mathcal{D}^{1,p}(\mathbb{R}^N)$. Using Proposition 7.2.3, we can assume that $u_n \to u$ almost everywhere and

$$\sum_{k=1}^{\infty} 2^{kp} \int_{\mathbb{R}^N} |\nabla(u_{k+1} - u_k)|^p dx < \infty.$$

We define

$$U_k = \{|u_{k+1} - u_k| > 2^{-k}\}, \quad \omega_m = \bigcup_{k=m}^{\infty} U_k.$$

Corollary 7.2.10 implies that for every $k$,

$$\text{cap}_p(U_k) \le 2^{kp} \int_{\mathbb{R}^N} |\nabla(u_{k+1} - u_k)|^p dx.$$

It follows from Corollary 7.1.10 that for every $m$,

$$\text{cap}_p(\omega_m) \le \sum_{k=m}^{\infty} 2^{kp} \int_{\mathbb{R}^N} |\nabla(u_{k+1} - u_k)|^p dx \to 0, \quad m \to \infty.$$

We obtain, for every $x \in \mathbb{R}^N \setminus \omega_m$ and every $k \ge j \ge m$,

$$|u_j(x) - u_k(x)| \le 2^{1-j},$$

so that $(u_n)$ converges simply to $v$ on $\mathbb{R}^N \setminus \bigcap_{m=1}^{\infty} \omega_m$. Moreover, $v\big|_{\mathbb{R}^N \setminus \omega_m}$ is continuous, since the convergence of $(u_n)$ on $\mathbb{R}^N \setminus \omega_m$ is uniform. For $x \in \bigcap_{m=1}^{\infty} \omega_m$, we define $v(x) = 0$. Since by Proposition 7.2.6, $m(\omega_m) \to 0$, we conclude that $u = v$ almost everywhere. □

## 7.3 Functions of Bounded Variations

A function is of *bounded variation* if its first-order derivatives, in the sense of distributions, are bounded measures.

**Definition 7.3.1** Let $\Omega$ be an open subset of $\mathbb{R}^N$. The divergence of $v \in C^1(\Omega; \mathbb{R}^N)$ is defined by

$$\text{div } v = \sum_{k=1}^{N} \partial_k v_k.$$

The total variation of $u \in L^1_{\text{loc}}(\Omega)$ is defined by

$$||Du||_\Omega = \sup \left\{ \int_\Omega u \text{ div } v \, dx : v \in \mathcal{D}(\Omega; \mathbb{R}^N), ||v||_\infty \leq 1 \right\},$$

where

$$||v||_\infty = \sup_{x \in \Omega} \left( \sum_{k=1}^{N} (v_k(x))^2 \right)^{1/2}.$$

**Theorem 7.3.2** *Let* $(u_n)$ *be such that* $u_n \to u$ *in* $L^1_{\text{loc}}(\Omega)$. *Then*

$$||Du||_\Omega \leq \varliminf_{n \to \infty} ||Du_n||_\Omega.$$

**Proof** Let $v \in \mathcal{D}(\Omega; \mathbb{R}^N)$ be such that $||v||_\infty \leq 1$. We have, by definition,

$$\int_\Omega u \text{ div } v \, dx = \lim_{n \to \infty} \int_\Omega u_n \text{ div } v \, dx \leq \varliminf_{n \to \infty} ||Du_n||_\Omega.$$

It suffices then to take the supremum with respect to $v$.                       $\square$

**Theorem 7.3.3** *Let* $u \in W^{1,1}_{\text{loc}}(\Omega)$. *Then the following properties are equivalent:*

*(a)* $\nabla u \in L^1(\Omega; \mathbb{R}^N)$;
*(b)* $||Du||_\Omega < \infty$.

*In this case,*

$$||Du||_\Omega = ||\nabla u||_{L^1(\Omega)}.$$

## *Proof*

(a)  Assume that $\nabla u \in L^1(\Omega; \mathbb{R}^N)$. Let $v \in \mathcal{D}(\Omega; \mathbb{R}^N)$ be such that $||v||_\infty \leq 1$. It
follows from the Cauchy–Schwarz inequality that

$$\int_\Omega u \text{ div } v \, dx = -\int_\Omega \sum_{k=1}^N v_k \partial_k u \, dx \leq \int_\Omega |\nabla u| dx.$$

Hence $||Du||_\Omega \leq ||\nabla u||_{L^1(\Omega)}$.

Theorem 4.3.11 implies the existence of $(w_n) \subset \mathcal{D}(\Omega; \mathbb{R}^N)$ converging to
$\nabla u$ in $L^1(\Omega; \mathbb{R}^N)$. We can assume that $w_n \to \nabla u$ almost everywhere on $\Omega$.
Let us define

$$v_n = w_n / \sqrt{|w_n|^2 + 1/n}.$$

We infer from Lebesgue's dominated convergence theorem that

$$||\nabla u||_{L^1(\Omega)} = \int_\Omega |\nabla u| dx = \lim_{n \to \infty} \int_\Omega v_n \cdot \nabla u \, dx \leq ||Du||_\Omega.$$

(b)  Assume that $||Du||_\Omega < \infty$, and define

$$\omega_n = \{x \in \Omega : d(x, \partial\Omega) > 1/n \text{ and } |x| < n\}.$$

Then by the preceding step, we obtain

$$||\nabla u||_{L^1(\omega_n)} = ||Du||_{\omega_n} \leq ||Du||_\Omega < \infty.$$

Levi's theorem ensures that $\nabla u \in L^1(\Omega; \mathbb{R}^N)$.                                  □

*Example*  There exists a function everywhere differentiable on $[-1, 1]$ such that
$||Du||_{]-1,1[} = +\infty$. We define

$$u(x) = 0, \qquad\qquad x = 0,$$
$$= x^2 \sin \tfrac{1}{x^2}, \qquad 0 < |x| \leq 1.$$

We obtain

$$u'(x) = 0, \qquad\qquad\qquad\qquad x = 0,$$
$$= 2x \sin \tfrac{1}{x^2} - \tfrac{2}{x} \cos \tfrac{1}{x^2}, \qquad 0 < |x| \leq 1.$$

The preceding theorem implies that

$$+\infty = \lim_{n\to\infty} ||u'||_{L^1(]1/n,1[)} \le ||Du||_{]-1,1[}.$$

Indeed,

$$2\int_0^1 |\cos\frac{1}{x^2}|\frac{dx}{x} = \int_1^\infty |\cos t|\frac{dt}{t} = +\infty.$$

The function $u$ has no weak derivative!

*Example (Cantor's Function)* There exists a continuous nondecreasing function with almost everywhere zero derivative and positive total variation. We use the notation of the last example of Sect. 2.2. We consider the Cantor set $C$ corresponding to $\ell_n = 1/3^{n+1}$. Observe that

$$m(C) = 1 - \sum_{j=0}^\infty 2^j/3^{j+1} = 0.$$

We define on $\mathbb{R}$,

$$u_n(x) = \left(\frac{3}{2}\right)^n \int_0^x \chi_{C_n}(t)dt.$$

It is easy to verify by symmetry that

$$||u_{n+1} - u_n||_\infty \le \frac{1}{3}\frac{1}{2^{n+1}}.$$

By the Weierstrass test, $(u_n)$ converges uniformly to the *Cantor's function* $u \in C(\mathbb{R})$. For $n \ge m$, $u'_n = 0$ on $\mathbb{R} \setminus C_m$. The closing lemma implies that $u' = 0$ on $\mathbb{R} \setminus C_m$. Since $m$ is arbitrary, $u' = 0$ on $\mathbb{R} \setminus C$. Theorems 7.3.2 and 7.3.3 ensure that

$$||Du||_\mathbb{R} \le \lim_{n\to\infty} ||u'_n||_{L^1(\mathbb{R})} = 1.$$

Let $v \in \mathcal{D}(\mathbb{R})$ be such that $||v||_\infty = 1$ and $v = -1$ on $[0, 1]$ and integrate by parts:

$$\int_\mathbb{R} v'u \, dx = \lim_{n\to\infty} \int_\mathbb{R} v'u_n \, dx = -\lim_{n\to\infty} \int_\mathbb{R} vu'_n dx = \lim_{n\to\infty} \left(\frac{3}{2}\right)^n m(C_n) = 1.$$

We conclude that $||Du||_\mathbb{R} = 1$. The function $u$ has no weak derivative.

**Definition 7.3.4** Let $\Omega$ be an open subset of $\mathbb{R}^N$. On the space

$$BV(\Omega) = \{u \in L^1(\Omega) : ||Du||_\Omega < \infty\},$$

we define the norm

$$||u||_{BV(\Omega)} = ||u||_{L^1(\Omega)} + ||Du||_\Omega$$

and the distance of strict convergence

$$d_S(u, v) = ||u - v||_{L^1(\Omega)} + \big|||Du||_\Omega - ||Dv||_\Omega\big|.$$

*Remark* It is clear that convergence in norm implies strict convergence.

*Example* The space $BV(]0, \pi[)$, with the distance of strict convergence, is not complete. We define on $]0, \pi[$,

$$u_n(x) = \frac{1}{n}\cos nx,$$

so that $u_n \to 0$ in $L^1(]0, \pi[)$. By Theorem 7.3.3, for every $n$,

$$||Du_n||_{]0,\pi[} = \int_0^\pi |\sin nx|dx = 2.$$

Hence $\lim\limits_{j,k\to\infty} d_S(u_j, u_k) = \lim\limits_{j,k\to\infty} ||u_j - u_k||_{L^1(]0,\pi[)} = 0$. If $\lim\limits_{n\to\infty} d_S(u_n, v) = 0$, then $v = 0$. But $\lim\limits_{n\to\infty} d_S(u_n, 0) = 2$. This is a contradiction.

**Proposition 7.3.5** *The normed space $BV(\Omega)$ is complete.*

**Proof** Let $(u_n)$ be a Cauchy sequence on the normed space $BV(\Omega)$. Then $(u_n)$ is a Cauchy sequence in $L^1(\Omega)$, so that $u_n \to u$ in $L^1(\Omega)$.

Let $\varepsilon > 0$. There exists $m$ such that for $j, k \geq m$, $||D(u_j - u_k)||_\Omega \leq \varepsilon$. Theorem 7.3.2 implies that for $k \geq m$, $||D(u_k - u)|| \leq \lim\limits_{j\to\infty} ||D(u_j - u_k)||_\Omega \leq \varepsilon$. Since $\varepsilon > 0$ is arbitrary, $||D(u_k - u)||_\Omega \to 0, k \to \infty$.  $\square$

**Lemma 7.3.6** *Let $u \in L^1_{\text{loc}}(\mathbb{R}^N)$ be such that $||Du||_{\mathbb{R}^N} < \infty$. Then*

$$||\nabla(\rho_n * u)||_{L^1(\mathbb{R}^N)} \leq ||Du||_{\mathbb{R}^N} \text{ and } ||Du||_{\mathbb{R}^N} = \lim\limits_{n\to\infty} ||\nabla(\rho_n * u)||_{L^1(\mathbb{R}^N)}.$$

**Proof** Let $v \in \mathcal{D}(\mathbb{R}^N; \mathbb{R}^N)$ be such that $||v||_\infty \leq 1$. It follows from Proposition 4.3.15 that

$$\int_{\mathbb{R}^N} (\rho_n * u) \operatorname{div} v \, dx = \int_{\mathbb{R}^N} u \sum_{k=1}^N \rho_n * \partial_k v_k dx = \int_{\mathbb{R}^N} u \sum_{k=1}^N \partial_k(\rho_n * v_k)dx.$$

The Cauchy–Schwarz inequality implies that for every $x \in \mathbb{R}^N$,

$$\sum_{k=1}^{N} (\rho_n * v_k(x))^2 = \sum_{k=1}^{N} \left( \int_{\mathbb{R}^N} \rho_n(x-y) v_k(y) dy \right)^2 \leq \sum_{k=1}^{N} \int_{\mathbb{R}^N} \rho_n(x-y)(v_k(y))^2 dy \leq 1.$$

Hence we obtain

$$\int_{\mathbb{R}^N} (\rho_n * u) \operatorname{div} v \, dx \leq ||Du||_{\mathbb{R}^N},$$

and by Theorem 7.3.3, $||\nabla(\rho_n * u)||_{L^1(\mathbb{R}^N)} \leq ||Du||_{\mathbb{R}^N}$.

By the regularization theorem, $\rho_n * u \to u$ in $L^1_{\text{loc}}(\mathbb{R}^N)$. Theorems 7.3.2 and 7.3.3 ensure that

$$||Du||_{\mathbb{R}^N} \leq \lim_{n \to \infty} ||\nabla(\rho_n * u)||_{L^1(\mathbb{R}^N)}. \qquad \square$$

**Theorem 7.3.7**

(a) *For every $u \in BV(\mathbb{R}^N)$, $(\rho_n * u)$ converges strictly to $u$.*
(b) *(Gagliardo–Nirenberg inequality.) Let $N \geq 2$. There exists $c_N > 0$ such that for every $u \in BV(\mathbb{R}^N)$,*

$$||u||_{N/(N-1)}(\mathbb{R}^N) \leq c_N ||Du||_{\mathbb{R}^N}.$$

*Proof*

(a) Proposition 4.3.14 and the preceding lemma imply the strict convergence of $(\rho_n * u)$ to $u$.
(b) Let $N \geq 2$. We can assume that $\rho_{n_k} * u \to u$ almost everywhere on $\mathbb{R}^N$. It follows from Fatou's lemma and Sobolev's inequality in $\mathcal{D}^{1,1}(\mathbb{R}^N)$ that

$$||u||_{N/(N-1)} \leq \lim_{k \to \infty} ||\rho_{n_k} * u||_{N/(N-1)} \leq c_N \lim_{n \to \infty} ||\nabla(\rho_{n_k} * u)||_1 = c_N ||Du||_{\mathbb{R}^N}.$$

$\square$

## 7.4 Perimeter

The *perimeter* of a smooth domain is the total variation of its characteristic function.

**Theorem 7.4.1** *Let $\Omega$ be an open subset of $\mathbb{R}^N$ of class $C^1$ with a bounded boundary $\Gamma$. Then*

$$\int_{\Gamma} d\gamma = ||D\chi_\Omega||_{\mathbb{R}^N}.$$

**Proof** Let $v \in \mathcal{D}(\mathbb{R}^N; \mathbb{R}^N)$ be such that $||v||_\infty \le 1$. The divergence theorem and the Cauchy–Schwarz inequality imply that

$$\int_\Omega \operatorname{div} v \, dx = \int_\Gamma v \cdot n \, d\gamma \le \int_\Gamma |v| \, |n| d\gamma \le \int_\Gamma d\gamma.$$

Taking the supremum with respect to $v$, we obtain $||D\chi_\Omega||_{\mathbb{R}^N} \le \int_\Gamma d\gamma$.

We use the notations of Definition 9.4.1 and define

$$U = \{x \in \mathbb{R}^N : \nabla\varphi(x) \ne 0\},$$

so that $\Gamma \subset U$. The theorem of partitions of unity ensures the existence of $\psi \in \mathcal{D}(U)$ such that $0 \le \psi \le 1$ and $\psi = 1$ on $\Gamma$. We define

$$\begin{aligned}
v(x) &= \psi(x)\nabla\varphi(x)/|\nabla\varphi(x)|, & x \in U\\
&= 0, & x \in \mathbb{R}^N \setminus U.
\end{aligned}$$

It is clear that $v \in \mathcal{K}(\mathbb{R}^N; \mathbb{R}^N)$, and for every $\gamma \in \Gamma$, $v(\gamma) = n(\gamma)$. For every $m \ge 1$, $w_m = \rho_m * v \in \mathcal{D}(\mathbb{R}^N; \mathbb{R}^N)$. We infer from the divergence and regularization theorems that

$$\lim_{m\to\infty} \int_\Omega \operatorname{div} w_m \, dx = \lim_{m\to\infty} \int_\Gamma w_m \cdot n \, d\gamma = \int_\Gamma n \cdot n \, d\gamma = \int_\Gamma d\gamma.$$

By definition, $||v||_\infty \le 1$, and by the Cauchy–Schwarz inequality,

$$\sum_{k=1}^N (\rho_m * v_k(x))^2 = \sum_{k=1}^N \left( \int_{\mathbb{R}^N} \rho_m(x-y)v_k(y)dy \right)^2 \le \sum_{k=1}^N \int_{\mathbb{R}^N} \rho_m(x-y)(v_k(y))^2 dy \le 1.$$

We conclude that $\int_\Gamma d\gamma \le ||D\chi_\Omega||_{\mathbb{R}^N}$. $\qquad\qquad\qquad\qquad\qquad\qquad\square$

The preceding theorem suggests a functional definition of the perimeter due to De Giorgi.

**Definition 7.4.2** Let $A$ be a measurable subset of $\mathbb{R}^N$. The perimeter of $A$ is defined by $p(A) = ||D\chi_A||_{\mathbb{R}^N}$.

**Definition 7.4.3** Let $N \ge 2$ and let $\Omega$ be an open subset of $\mathbb{R}^N$. The Cheeger constant of $\Omega$ is defined by

$$h(\Omega) = \inf\{p(\omega)/m(\omega) : \omega \subset\subset \Omega \text{ and } \omega \text{ is of class } C^1\}.$$

*Example* Let $\Omega = B(0,1) \subset \mathbb{R}^N$. For every $0 < r < 1, h(\Omega) \leq N/r$, so that $h(\Omega) \leq N$. Assume that $\omega \subset\subset \Omega$ is of class $C^1$. The divergence theorem, applied to the vector field $v(x) = x$, implies that

$$Nm(\omega) = \int_\omega \operatorname{div} v \, dx = \int_{\partial\omega} v.n \, d\gamma \leq \int_{\partial\omega} d\gamma = p(\omega).$$

We conclude that $h(\Omega) = N$.

**Theorem 7.4.4 (S.T. Yau, 1975)** *Let $\Omega$ be an open subset of $\mathbb{R}^N$. Then*

$$h(\Omega) = \inf_{\substack{u \in \mathcal{D}(\Omega) \\ u \neq 0}} \frac{\displaystyle\int_\Omega |\nabla u| dx}{\displaystyle\int_\Omega |u| dx}.$$

*Proof*

(a)  Let $u \in \mathcal{D}(\Omega)$. Using Cavalieri principle (Corollary 2.2.34), the Morse–Sard theorem (Theorem 9.3.1), and the coarea formula (Theorem 9.3.3), we obtain

$$h(\Omega) \int_\Omega |u| dx = h(\Omega) \left[ \int_0^\infty m(\{u > t\}) dt + \int_{-\infty}^0 m(\{u < t\}) dt \right]$$

$$\leq \int_0^\infty dt \int_{u=t} d\gamma + \int_{-\infty}^0 dt \int_{u=t} d\gamma = \int_\Omega |\nabla u| dx,$$

so that

$$h(\Omega) \leq c = \inf_{\substack{u \in \mathcal{D}(\Omega) \\ u \neq 0}} \frac{\displaystyle\int_\Omega |\nabla u| dx}{\displaystyle\int_\Omega |u| dx}.$$

(b)  Let $\omega \subset\subset \Omega$ be of class $C^1$. For $n$ large enough, $u_n = \rho_n * \chi_\omega \in \mathcal{D}(\Omega)$. Proposition 4.3.14 and Lemma 7.3.6 imply that, as $n \to \infty$,

$$\|u_n\|_1 \to \|\chi_\omega\|_1 = m(\omega), \|\nabla u_n\|_1 \to \|D\chi_\omega\|_{\mathbb{R}^N} = p(\omega).$$

We conclude that

$$\frac{p(\omega)}{m(\omega)} = \lim_{n\to\infty} \frac{\displaystyle\int_\Omega |\nabla u_n| dx}{\displaystyle\int_\Omega |u_n| dx} \geq c,$$

and $h(\Omega) \geq c$.                                                                                □

**Corollary 7.4.5 (J. Cheeger, 1970)** *Let $\Omega$ be an open subset of $\mathbb{R}^N$. Then*

$$\frac{h^2(\Omega)}{4} \leq \inf_{\substack{u \in \mathcal{D}(\Omega) \\ u \neq 0}} \frac{\displaystyle\int_\Omega |\nabla u|^2 dx}{\displaystyle\int_\Omega u^2 dx}.$$

***Proof*** Let $u \in \mathcal{D}(\Omega)$. By the preceding theorem, the function $v = u^2$ satisfies

$$h(\Omega) \int_\Omega |v| dx \leq \int_\Omega |\nabla v| dx.$$

The Cauchy-Schwarz inequality implies that

$$h(\Omega) \int_\Omega u^2 dx \leq \int_\Omega 2|u| \, |\nabla u| dx \leq 2 \left( \int_\Omega u^2 dx \right)^{1/2} \left( \int_\Omega |\nabla u|^2 dx \right)^{1/2}. \qquad \square$$

**Lemma 7.4.6** *Let $1 \leq p < N$, let $K$ be a compact subset of $\mathbb{R}^N$, and $a > \mathrm{cap}_p(K)$. Then there exist $V$ open and $v \in \mathcal{D}(\mathbb{R}^N)$ such that $K \subset V$, $\chi_V \leq v$, and*
$$\int_\Omega |\nabla v| \, dx < a.$$

***Proof*** By assumption, there exist $u \in \mathcal{D}^{1,p}(\mathbb{R}^N)$ and $U$ open such that $K \subset U$, $\chi_U \leq u$, and

$$\int_{\mathbb{R}^N} |\nabla u|^p dx < a.$$

There exists $V$ open such that $K \subset V \subset\subset U$. For $m$ large enough, $\chi_V \leq w = \rho_m * u$ and

$$\int_{\mathbb{R}^N} |\nabla w|^p dx < a.$$

Let $\theta_n(x) = \theta(|x|/n)$ be a truncating sequence. For $n$ large enough, $\chi_V \leq v = \theta_n w$ and

$$\int_{\mathbb{R}^N} |\nabla v|^p dx < a. \qquad \square$$

**Theorem 7.4.7** *Let $N \geq 2$ and let $K$ be a compact subset of $\mathbb{R}^N$. Then*

$$\mathrm{cap}_1(K) = \inf\{p(U) : U \text{ is open and bounded, and } U \supset K\}.$$

**Proof** We denote by $\mathrm{Cap}_1(K)$ the second member of the preceding equality. Let $U$ be open, bounded, and such that $U \supset K$. Define $u_n = \rho_n * \chi_U$. For $n$ large enough, $u \in \mathcal{D}_K^{1,1}(\mathbb{R}^N)$. Lemma 7.3.6 implies that for $n$ large enough,

$$\mathrm{cap}_1(K) \leq \int_{\mathbb{R}^N} |\nabla u_n| dx \leq ||D\chi_U||_{\mathbb{R}^N} = p(U).$$

Taking the infimum with respect to $U$, we obtain $\mathrm{cap}_1(K) \leq \mathrm{Cap}_1(K)$.

Let $a > \mathrm{cap}_1(K)$. By the preceding lemma, there exist $V$ open and $v \in \mathcal{D}(\mathbb{R}^N)$ such that $K \subset V$, $\chi_V \leq v$, and $\int_{\mathbb{R}^N} |\nabla v| dx < a$. We deduce from the Morse–Sard theorem and from the coarea formula that

$$\mathrm{Cap}_1(K) \leq \int_0^1 dt \int_{v=t} d\gamma \leq \int_0^\infty dt \int_{v=t} d\gamma = \int_{\mathbb{R}^N} |\nabla v| dx < a.$$

Since $a > \mathrm{cap}_1(K)$ is arbitrary, we conclude that $\mathrm{Cap}_1(K) \leq \mathrm{cap}_1(K)$.  □

## 7.5   Distribution Theory

> La mathématique est l'art de donner le même nom à des choses différentes.
>
> *Henri Poincaré*

> La mathématique est la science des choses qui se réduisent à leur définition.
>
> *Paul Valéry*

Distribution theory is a general framework including locally integrable functions. A *distribution* is a continuous linear functional on the space of test functions. Every distribution is infinitely differentiable, and differentiation of distributions is a continuous operation. We denote by $\Omega$ an open subset of $\mathbb{R}^N$.

**Definition 7.5.1** A sequence $(u_n)$ converges to $u$ in $\mathcal{D}(\Omega)$ if there exists a compact subset $K$ of $\Omega$ such that for every $n$, spt $u_n \subset K$, and if for every $\alpha \in \mathbb{N}^N$,

$$||\partial^\alpha(u_n - u)||_\infty \to 0, n \to \infty.$$

**Definition 7.5.2** A distribution on $\Omega$ is a linear functional $f : \mathcal{D}(\Omega) \to \mathbb{R}$ such that for every sequence $(u_n)$ converging to $u$ in $\mathcal{D}(\Omega)$, $\langle f, u_n \rangle \to \langle f, u \rangle, n \to \infty$. We denote by $\mathcal{D}^*(\Omega)$ the space of distributions on $\Omega$.

*Example* The distribution corresponding to $f \in L^1_{loc}(\Omega)$ is defined on $\mathcal{D}(\Omega)$ by

$$\langle f, u \rangle = \int_\Omega f u \, dx.$$

By the annulation theorem, the *functional* $f : \mathcal{D}(\Omega) \to \mathbb{R}$ characterizes the *function* $f \in L^1_{loc}(\Omega)$. Assume that $u_n \to u$ in $\mathcal{D}(\Omega)$. Then there exists a compact subset $K$ of $\Omega$ such that, for every $n$, spt $u_n \subset K$. Hence we obtain

$$\left| \int_\Omega f u_n dx - \int_\Omega f u dx \right| \leq \int_\Omega |f| \, |u_n - u| dx \leq \int_K |f| dx \|u_n - u\|_\infty \to 0, n \to \infty.$$

**Definition 7.5.3** Let $f \in \mathcal{D}^*(\Omega)$ and $\alpha \in \mathbb{N}^N$. The derivative of order $\alpha$ of $f$ (in the sense of distributions) is defined on $\mathcal{D}(\Omega)$ by

$$\langle \partial^\alpha f, u \rangle = (-1)^{|\alpha|} \langle f, \partial^\alpha u \rangle.$$

It is easy to verify that $\partial^\alpha f \in \mathcal{D}^*(\Omega)$.

*Examples*

(a) If $g = \partial^\alpha f$ in the *weak sense*, then $g = \partial^\alpha f$ in the *sense of distributions*. Indeed, for every $u \in \mathcal{D}(\Omega)$,

$$\langle \partial^\alpha f, u \rangle = (-1)^{|\alpha|} \langle f, \partial^\alpha u \rangle = (-1)^{|\alpha|} \int_\Omega f \partial^\alpha u dx = \int_\Omega g u dx = \langle g, u \rangle.$$

(b) The everywhere derivable function

$$f(x) = 0, \qquad\qquad x = 0,$$
$$= x^2 \sin \frac{1}{x^2}, \qquad 0 < |x| < 1,$$

has a classical derivative $f'$ and a derivative in the sense of distributions

$$\mathcal{D}(-1, 1[) \to \mathbb{R} : u \mapsto -\int_{-1}^1 f u' dx.$$

Those two objects are *different* since, for every $\varepsilon > 0$, $\int_{-\varepsilon}^\varepsilon |f'| dx = +\infty$.

**Definition 7.5.4** The sequence $(f_n)$ converges to $f$ in $\mathcal{D}^*(\Omega)$ if for every $u \in \mathcal{D}(\Omega)$, $\langle f_n, u \rangle \to \langle f, u \rangle, n \to \infty$.

*Example* If $f_n \to f$ in $L^1_{\text{loc}}(\Omega)$, then $f_n \to f$ in $\mathcal{D}^*(\Omega)$. Indeed, for every $u \in \mathcal{D}(\Omega)$

$$\left| \int_\Omega f_n u \, dx - \int_\Omega f u \, dx \right| \leq \int_\Omega |f_n - f| \, |u| \, dx \leq \|u\|_\infty \int_{\text{spt } u} |f_n - f| \, dx \to 0, \, n \to \infty.$$

**Theorem 7.5.5** *Let $\alpha \in \mathbb{N}^N$ and let $(f_n)$ be a sequence converging to $f$ in $\mathcal{D}^*(\Omega)$. Then $(\partial^\alpha f_n)$ converges to $\partial^\alpha f$ in $\mathcal{D}^*(\Omega)$.*

**Proof** For every $u \in \mathcal{D}(\Omega)$, we have

$$\langle \partial^\alpha f, u \rangle = (-1)^{|\alpha|} \langle f, \partial^\alpha u \rangle = \lim_{n \to \infty} (-1)^{|\alpha|} \langle f_n, \partial^\alpha u \rangle = \lim_{n \to \infty} \langle \partial^\alpha f_n, u \rangle. \qquad \square$$

We now prove a variant of the Banach–Steinhaus theorem.

**Theorem 7.5.6** *Let $(f_j) \subset \mathcal{D}^*(\Omega)$ be a sequence converging simply to the functional $f : \mathcal{D}(\Omega) \to \mathbb{R}$. Then $f \in \mathcal{D}^*(\Omega)$, so that $f_j \to f$ in $\mathcal{D}^*(\Omega)$.*

**Proof** The linearity of $f$ is clear. Assume, for the sake of obtaining a contradiction, that there exists $(u_n) \subset \mathcal{D}(\Omega)$ such that $u_n \to 0$ in $\mathcal{D}(\Omega)$ and $\varlimsup_{n \to \infty} |f(u_n)| > 0$. We can assume that $\varlimsup_{n \to \infty} f(u_n) > 0$. Using Cantor's diagonal argument, we construct a subsequence $(v_k)$ of $(u_n)$ such that for every $k$ and every $|\alpha| \leq k$,

$$0 < c < f(v_k), \quad \|\partial^\alpha v_k\|_\infty \leq 1/2^k.$$

We choose $v_{k_1} = v_1$ and $f_{j_1}$ such that $c < \langle f_{j_1}, v_{k_1} \rangle$. Given $v_{k_1}, \ldots, v_{k_{n-1}}$ and $f_{j_1}, \ldots, f_{j_{n-1}}$, there exists $v_{k_n}$ such that for $m \leq n - 1$,

$$|\langle f_{j_m}, v_{k_n} \rangle| \leq 1/2^{n-m}.$$

There also exists $f_{j_n}$ such that

$$nc < \sum_{m=1}^{n} \langle f_{j_n}, v_{k_m} \rangle.$$

By the Weierstrass test, $\sum_{m=1}^{\infty} v_{k_m} = w$ in $\mathcal{D}(\Omega)$. Hence we obtain, for every $n$,

$$\langle f_{j_n}, w \rangle = \sum_{m=1}^{\infty} \langle f_{j_n}, v_{k_m} \rangle > nc - \sum_{m=n+1}^{\infty} 1/2^{m-n} = nc - 1.$$

But then $\langle f_{j_n}, w \rangle \to +\infty, n \to \infty$. This is a contradiction. $\qquad \square$

The preceding theorem explains why every natural linear functional defined on $\mathcal{D}(\Omega)$ is continuous.

Distributions also generalize positive measures and bounded measures.

**Theorem 7.5.7** *Let* $f: \mathcal{D}(\Omega) \to \mathbb{R}$ *be a linear functional such that* $\langle f, u \rangle \geq 0$ *when* $u \geq 0$. *Then* $f$ *is a distribution and the restriction to* $\mathcal{D}(\Omega)$ *of a positive measure* $\mu: \mathcal{K}(\Omega) \to \mathbb{R}$.

**Proof** Let $\omega \subset\subset \Omega$. By the theorem of partitions of unity, there exists $\psi \in \mathcal{D}(\Omega)$ such that $0 \leq \psi \leq 1$ and $\psi = 1$ on $\omega$. For every $u \in \mathcal{D}(\Omega)$ such that spt $u \subset \omega$, we have

$$\langle f, u \rangle \leq \langle f, \|u\|_\infty \psi \rangle = c_\omega \|u\|_\infty.$$

Hence $f$ is a distribution.

Let $v \in \mathcal{K}(\Omega)$ be such that spt $v \subset \omega$, and define $v_n = \rho_n * v$. For $n$ large enough, spt $v_n \subset \omega$. The regularization theorem implies that

$$\lim_{j,k\to\infty} |\langle f, v_j \rangle - \langle f, v_k \rangle| \leq c_\omega \lim_{j,k\to\infty} \|v_j - v_k\|_\infty = 0.$$

We define

$$\langle \mu, v \rangle = \lim_{n\to\infty} \langle f, v_n \rangle. \qquad \square$$

**Lemma 7.5.8** *Let* $f: \mathcal{D}(\Omega; \mathbb{R}^M) \to \mathbb{R}$ *be a linear functional such that*

$$c_f = \sup\{\langle f, v \rangle : v \in \mathcal{D}(\Omega; \mathbb{R}^M), \|v\|_\infty \leq 1\} < +\infty.$$

*Then* $f$ *is the restriction to* $\mathcal{D}(\Omega; \mathbb{R}^M)$ *of a finite measure* $\mu: \mathcal{K}(\Omega; \mathbb{R}^M) \to \mathbb{R}$ *such that* $c_f = \|\mu\|_\Omega = \int_\Omega d|\mu|$, *and there exists* $g: \Omega \to \mathbb{R}^M$ *satisfying:*

*(a)* $g$ *is* $|\mu|$-*measurable;*
*(b)* $|g(x)| = 1$, $|\mu|$-*almost everywhere on* $\Omega$;
*(c)* *for all* $v \in \mathcal{D}(\Omega; \mathbb{R}^M)$, $\langle f, v \rangle = \int_\Omega v \cdot g \, d|\mu|$.

**Proof** Let $v \in \mathcal{K}(\Omega; \mathbb{R}^M)$. For $n$ large enough, spt $\rho_n * v \subset\subset \Omega$. The regularization theorem implies that

$$\lim_{j,k\to\infty} |\langle f, \rho_j * v \rangle - \langle f, \rho_k * v \rangle| \leq c_f \lim_{j,k\to\infty} \|\rho_j * v - \rho_k * v\|_\infty = 0.$$

We define

$$\langle \mu, v \rangle = \lim_{n\to\infty} \langle f, \rho_n * v \rangle.$$

Since

$$\langle \mu, v \rangle \leq c_f \lim_{n \to \infty} \|\rho_n * v\|_\infty = c_f \|v\|_\infty,$$

the functional $\mu \colon \mathcal{K}(\Omega; \mathbb{R}^M) \to \mathbb{R}$ is a bounded measure such that

$$\|\mu\|_\Omega = \sup\{\langle \mu, v \rangle \ : \ v \in \mathcal{K}(\Omega; \mathbb{R}^M), \|v\|_\infty \leq 1\} \leq c_f.$$

We conclude that $\|\mu\|_\Omega = c_f$.

Let $(\psi_n) \subset \mathcal{D}(\Omega)$ be given by Proposition 6.1.16 so that $\psi_n \geq 0$ and, for every $\omega \subset\subset \Omega$, $\sum_{n=1}^{m_\omega} \psi_n = 1$ on $\omega$. Let us recall that

$$\int_\Omega \sum_{n=1}^{m} \psi_n \, d|\mu| = \sup\{\langle \mu, v \rangle \colon v \in \mathcal{K}(\Omega; \mathbb{R}^M), |v| \leq \sum_{n=1}^{m} \psi_n\} \leq \|\mu\|_\Omega.$$

Using Levi's theorem, we obtain

$$\int_\Omega d|\mu| = \int_\Omega \sum_{\nu=1}^{\infty} \psi_n \, d|\mu| = \sup\{\langle \mu, v \rangle \colon v \in \mathcal{K}(\Omega; \mathbb{R}^M), \|v\|_\infty \leq 1\} = \|\mu\|_\Omega.$$

Finally the existence of $g \colon \Omega \to \mathbb{R}^M$ satisfying (a), (b), and (c) follows from Theorem 5.3.14.                                                                                                     □

**Theorem 7.5.9**  *Let $u \in L^1_{loc}(\Omega)$ be such that*

$$\|Du\|_\Omega = \sup\{\int_\Omega u \operatorname{div} v dx \colon v \in \mathcal{D}(\Omega; \mathbb{R}^N), \|v\|_\infty \leq 1\} < +\infty.$$

*Then*

$$f \colon \mathcal{D}(\Omega; \mathbb{R}^N) \to \mathbb{R} \colon v \mapsto \sum_{j=1}^{N} \langle \partial_j u, v_j \rangle$$

*is the restriction to $\mathcal{D}(\Omega; \mathbb{R}^N)$ of a finite measure $Du \colon \mathcal{K}(\Omega; \mathbb{R}^N) \to \mathbb{R}$ such that $\|Du\|_\Omega = \int_\Omega d|Du|$, and there exists $g \colon \Omega \to \mathbb{R}^N$ satisfying:*

*(a) $g$ is $|Du|$-measurable;*
*(b) $|g(x)| = 1$, $|Du|$-almost everywhere on $\Omega$;*
*(c) for all $v \in \mathcal{D}(\Omega; \mathbb{R}^N)$, $\sum_{j=1}^{N} \langle \partial_j u, v_j \rangle = \int_\Omega v \cdot g \, d|Du|.$*

**Proof** Since, for every $v \in \mathcal{D}(\Omega; \mathbb{R}^N)$,

$$\langle f, v \rangle = \sum_{j=1}^{N} \langle \partial_j u, v_j \rangle = - \int_{\Omega} u \, \mathrm{div} v dx,$$

it suffices to use the preceding lemma. □

The next result improves Theorem 7.4.1.

**Theorem 7.5.10** *Let $\Omega$ be an open subset of $\mathbb{R}^N$ of class $C^1$ with a bounded boundary $\Gamma$, and let $v \in \mathcal{K}(\mathbb{R}^N; \mathbb{R}^N)$. Then*

$$\langle D\chi_\Omega, v \rangle = - \int_{\Gamma} v \cdot n \, d\gamma.$$

**Proof** The regularization theorem and the divergence theorem imply that

$$\langle D\chi_\Omega, v \rangle = \lim_{m \to \infty} \sum_{j=1}^{N} \langle \partial_j \chi_\Omega, \rho_m * v \rangle$$

$$= \lim_{m \to \infty} - \sum_{j=1}^{N} \int_{\Omega} \partial_j (\rho_m * v) dx$$

$$= \lim_{m \to \infty} - \int_{\Gamma} \rho_m * v \cdot n \, d\gamma$$

$$= - \int_{\Gamma} v \cdot n \, d\gamma. \qquad \Box$$

**Theorem 7.5.11** *(Density theorem in $BV(\Omega)$). Let $u \in BV(\Omega)$. Then there exists $(u_n) \subset C^\infty(\Omega) \cap W^{1,1}(\Omega)$ such that*

$$\lim_{n \to \infty} \|u_n - u\|_{L^1(\Omega)} = 0, \quad \lim_{n \to \infty} \int_{\Omega} |\nabla u_n| dx = \|Du\|_\Omega.$$

**Proof** Let us first prove that, for every $\varepsilon > 0$, there exists $v \in C^\infty(\Omega) \cap W^{1,1}(\Omega)$ such that

$$\|v - u\|_{L^1(\Omega)} \le \varepsilon, \quad \int_{\Omega} |\nabla v| dx \le \|Du\|_\Omega + \varepsilon.$$

Let $(U_n)$ and $(\psi_n)$ be given by Proposition 6.1.16. Since $\psi_n \in \mathcal{D}(U_n)$, there exists, for every $n \ge 1$, $k_n$ such that

$$v_n = \rho_{k_n} * (\psi_n u) \in \mathcal{D}(U_n)$$

and

$$||v_n - \psi_n u||_{L^1(\Omega)} + ||\rho_{k_n} * (u\nabla\psi_n) - u\nabla\psi_n||_{L^1(\Omega)} \le \varepsilon/2^n. \qquad (*)$$

On every $\omega \subset\subset \Omega$, we have that

$$\sum_{n=1}^{\infty} v_n = \sum_{n=1}^{m_\omega} v_n \in C^\infty(\omega).$$

Hence

$$v = \sum_{n=1}^{\infty} v_n \in C^\infty(\Omega).$$

Moreover $(*)$ implies that

$$||v - u||_{L^1(\Omega)} = ||\sum_{n=1}^{\infty}(v_n - \psi_n u)||_{L^1(\Omega)} \le \varepsilon.$$

We deduce from Proposition 4.3.6 and Theorem 7.5.9 (c) that, for every $x \in \Omega$,

$$\nabla v_n(x) = \int_\Omega \nabla_x \rho_{k_n}(x - y)\psi_n(y)u(y)dy$$

$$= -\int_\Omega \nabla_y \rho_{k_n}(x - y)\psi_n(y)u(y)dy$$

$$= \int_\Omega \rho_{k_n}(x - y)\psi_n(y)g(y)d|Du| + \int_\Omega \rho_{k_n}(x - y)u(y)\nabla\psi_n(y)dy$$

$$= \int_\Omega \rho_{k_n}(x - y)\psi_n(y)g(y)d|Du| + \rho_{k_n} * (u\nabla\psi_n)(x).$$

It follows from Fubini's theorem and Theorem 7.5.9 (b) that

$$\int_\Omega dx \left| \int_\Omega \rho_{k_n}(x - y)\psi_n(y)g(y)d|Du| \right| \le \int_\Omega dx \int_\Omega \rho_{k_n}(x - y)\psi_n(y)d|Du|$$

$$= \int_\Omega d|Du| \int_\Omega \rho_{k_n}(x - y)\psi_n(y)dx$$

$$= \int_\Omega \psi_n d|Du|. \qquad (**)$$

It is clear that, for every $x \in \Omega$,

$$\nabla v(x) = \sum_{n=1}^{\infty} \nabla v_n(x)$$

$$= \sum_{n=1}^{\infty} \int_{\Omega} \rho_{k_n}(x-y)\psi_n(y)g(y)d|Du| + \sum_{n=1}^{\infty} \rho_{k_n} * (u\nabla\psi_n)(x)$$

$$= \sum_{n=1}^{\infty} \int_{\Omega} \rho_{k_n}(x-y)\psi_n(y)g(y)d|Du| + \sum_{n=1}^{\infty} \left[\rho_{k_n} * (u\nabla\psi_n) - u\nabla\psi_n\right](x).$$

Using $(*)$ and $(**)$, we conclude from Levi's theorem that

$$\int_{\Omega} |\nabla v|dx \le \sum_{n=1}^{\infty} \int_{\Omega} \psi_n d|Du| + \sum_{n=1}^{\infty} \varepsilon/2^n = \int_{\Omega} \sum_{n=1}^{\infty} \psi_n d|Du| + \varepsilon = \|Du\|_{\Omega} + \varepsilon.$$

The first part of the proof implies the existence of a sequence $(u_n) \subset C^{\infty}(\Omega) \cap W^{1,1}(\Omega)$ such that, for every $n$,

$$\|u_n - u\|_{L^1(\Omega)} \le 1/n, \quad \int_{\Omega} |\nabla u_n|dx \le \|Du\|_{\Omega} + 1/n.$$

Using Theorems 7.3.2 and 7.3.3, we obtain

$$\|Du\|_{\Omega} \le \varliminf_{n\to\infty} \int_{\Omega} |\nabla u_n|dx \le \varlimsup_{n\to\infty} \int_{\Omega} |\nabla u_n|dx \le \|Du\|_{\Omega}. \qquad \square$$

We shall prove various representation theorems.

*Notation.* Let $1 \le p < +\infty$ and let $\Omega$ be an open subset of $\mathbb{R}^N$. On $L^p(\Omega; \mathbb{R}^M)$, we define the norm

$$\|v\|_p \equiv \left(\int_{\Omega} |v|^p dx\right)^{1/p} = \left(\int_{\Omega} \left(\sum_{k=1}^{M} v_k^2\right)^{p/2} dx\right)^{1/p}.$$

Let us recall that $1/p + 1/p' = 1$.

**Lemma 7.5.12** *Let $1 < p < +\infty$ and let $g \in \left(L^p(\Omega; \mathbb{R}^M)\right)^*$. Then there exists one and only one $h \in L^{p'}(\Omega; \mathbb{R}^M)$ such that, for every $v \in L^p(\Omega; \mathbb{R}^M)$,*

$$\langle g, v \rangle = \int_{\Omega} h.v \, dx. \qquad (*)$$

*Moreover $\|g\|_{(L^p)^*} = \|h\|_{p'}$.*

**Proof** By the Riesz's representation theorem, there exists one and only one $h \in L^{p'}(\Omega; \mathbb{R}^M)$ satisfying $(*)$. Moreover

$$\|g\|_{(L^p)^*} = \sup_{\|v\|_p=1} \int_\Omega h \cdot v \; dx = \|h\|_{p'}. \qquad\qquad \square$$

**Theorem 7.5.13** *Let* $1 < p < +\infty$ *and let* $f \in \mathcal{D}^*(\Omega)$. *The following properties are equivalent:*

*(a)  there exists* $h \in L^{p'}(\Omega; \mathbb{R}^N)$ *such that* div $h = f$;
*(b)  * $\|f\|_* = \sup\{\langle f, u \rangle : u \in \mathcal{D}(\Omega), \|\nabla u\|_p \le 1\} < +\infty$.

*Moreover*

$$\|f\|_* = \min\Big\{\|h\|_{p'} : h \in L^{p'}(\Omega; \mathbb{R}^N), \text{div } h = f\Big\}.$$

***Proof*** If $f$ satisfies (a), it follows from Hölder's inequality that, on $\mathcal{D}(\Omega)$,

$$\langle f, u \rangle = \langle \text{div } h, u \rangle = -\int_\Omega h \cdot \nabla u \; dx \le \|h\|_{p'} \|\nabla u\|_p.$$

Hence $\|f\|_* \le \|h\|_{p'}$.

Assume that $f$ satisfies (b). We define $Y = \big\{\nabla u : u \in \mathcal{D}(\Omega)\big\}$. Since

$$A : \mathcal{D}(\Omega) \to Y : u \to \nabla u$$

is bijective, the Hahn–Banach theorem implies the existence of $g \in \big(L^p(\Omega; \mathbb{R}^N)\big)^*$ such that, for every $u \in \mathcal{D}(\Omega)$, $\langle g, \nabla u \rangle = \langle f, u \rangle$ and

$$\|g\|_{(L^p)^*} = \|f\|_*.$$

By the preceding lemma, there exists $h \in L^{p'}(\Omega; \mathbb{R}^N)$ such that, for every $v \in L^p(\Omega; \mathbb{R}^N)$, $\langle g, v \rangle = -\int_\Omega h \cdot v \; dx$ and

$$\|h\|_{p'} = \|g\|_{(L^p)^*}.$$

We conclude that $\|h\|_{p'} = \|f\|_*$ and that, for every $u \in \mathcal{D}(\Omega)$,

$$\langle f, u \rangle = \langle g, \nabla u \rangle = -\int_\Omega h \cdot \nabla u \; dx = \langle \text{div } h, u \rangle. \qquad\qquad \square$$

We now state the *representation theorem of L. Schwartz.*

**Theorem 7.5.14** *Let* $f \in \mathcal{D}^*(\Omega)$, *and let* $\omega \subset\subset \Omega$ *be the product of* $N$ *open intervals. Then there exist* $g \in C_0(\omega)$ *and* $\beta \in \mathbb{N}^N$ *such that* $f = \partial^\beta g$ *on* $\mathcal{D}(\omega)$.

**Lemma 7.5.15** *Let* $f \in \mathcal{D}^*(\Omega)$. *Then there exist* $\alpha \in \mathbb{N}^N$ *and* $c \ge 0$ *such that for all* $u \in \mathcal{D}(\omega)$,

$$|\langle f, u \rangle| \le c\|\partial^\alpha u\|_\infty.$$

**Proof**  By the fundamental theorem of calculus, for every $n \geq 1$, there exists $c_n > 0$ such that for all $u \in \mathcal{D}(\omega)$,

$$\sup_{|\alpha| \leq n} \|\partial^\alpha u\|_\infty \leq c_n \|\partial^{(n,\ldots,n)} u\|_\infty.$$

Assume, to obtain a contradiction, that for every $n \geq 1$, there exists $u_n \in \mathcal{D}(\omega)$ such that

$$n c_n \|\partial^{(n,\ldots,n)} u_n\|_\infty < \langle f, u_n \rangle.$$

Define $v_n = u_n / (n c_n \|\partial^{(n,\ldots,n)} u_n\|_\infty)$. Since for every $n \geq |\alpha|$, $\|\partial^\alpha v_n\|_\infty \leq 1/n$, we conclude that $v_n \to 0$ in $\mathcal{D}(\Omega)$ and $\langle f, v_n \rangle \to 0$. But this is impossible, since for every $n$, $\langle f, v_n \rangle > 1$.  $\square$

We prove *the existence of primitives of a distribution.*

**Lemma 7.5.16**  *Let $f \in \mathcal{D}^*(\omega)$, $1 \leq k \leq N$, $\gamma \in \mathbb{N}^N$, and $c \geq 0$ be such that for all $u \in \mathcal{D}(\omega)$,*

$$|\langle f, u \rangle| \leq c \|\partial_k \partial^\gamma u\|_\infty.$$

*Then there exist $F \in \mathcal{D}^*(\omega)$ and $C \geq 0$ such that $f = \partial_k F$ and for all $u \in \mathcal{D}(\omega)$,*

$$|\langle F, u \rangle \leq C \|\partial^\gamma u\|_\infty.$$

**Proof**  We can assume that $\omega = \,]0, 1[^N$ and $k = N$. Let $\varphi \in \mathcal{D}\left(]0, 1[\right)$ be such that $\int_0^1 \varphi ds = 1$. For every $u \in \mathcal{D}(\omega)$, there exists one and only one $v \in \mathcal{D}(\omega)$ such that

$$u(x) = \int_0^1 u(x', s) ds \, \varphi(x_N) + \partial_N v(x).$$

The function $v$ is given by the formula

$$v(x) = \int_0^{x_N} \left[ u(x', t) - \int_0^1 u(x', s) ds \, \varphi(t) \right] dt.$$

The distribution $F$ is defined by the formula

$$\langle F, u \rangle = -\langle f, v \rangle.$$

Since $\|\partial_N \partial^\gamma v\|_\infty \leq d \|\partial^\gamma u\|_\infty$, it is easy to finish the proof.  $\square$

Let us define

$$k(x, y) = -(1 - y)x, \ 0 \le x \le y \le 1,$$
$$= -(1 - x)y, \ 0 \le y \le x \le 1,$$

and

$$K(x, y) = \prod_{n=1}^{N} k(x_n, y_n).$$

**Lemma 7.5.17** *For every $u \in \mathcal{D}\left(]0, 1[^{N}\right)$, we have that*

$$u(x) = \int_{]0, 1[^{N}} K(x, y)\partial^{(2,\ldots,2)} u(y)dy.$$

**Proof** When $N = 1$, it suffices to integrate by parts. When $N \ge 2$, the result follows by induction from Fubini's theorem.                                                                          □

We now prove *the representation theorem of A. Pietsch* (1960).

**Lemma 7.5.18** *Let $\mu$ be a finite measure on $\omega$. Then there exists $g \in C_0(\omega)$ such that $\mu = \partial^{(2,\ldots,2)} g$ on $\mathcal{D}(\omega)$.*

**Proof** We can assume that $\omega = \ ]0, 1[^{N}$. By assumption, for every $u \in C_0(\omega)$,

$$\left|\langle \mu, u \rangle\right| \le c\|u\|_{\infty}, \qquad\qquad\qquad (*)$$

where $c = \|\mu\|_{\omega}$.

Let $u \in \mathcal{D}(\omega)$. By the preceding lemma, we have that

$$\langle \mu, u \rangle = \langle \mu, \int_{\omega} K(x, y)\partial^{(2,\ldots,2)} u(y)dy \rangle.$$

We shall prove that

$$\langle \mu, u \rangle = \int_{\omega} g(y)\partial^{(2,\ldots,2)} u(y)dy,$$

where

$$g(y) = \langle \mu, K(\cdot, y) \rangle.$$

Since

$$\left|K(x, y) - K(x, z)\right| \le \sum_{i=1}^{N} |y_j - z_j|,$$

it follows from (∗) that

$$|g(y) - g(z)| \leq c \sum_{j=1}^{N} |y_j - z_j|.$$

It is clear by definition that $g = 0$ on $\partial\omega$.

Define $v = \partial^{(2,\dots,2)} u$. The preceding lemma implies that

$$\left\| u(x) - 2^{-jN} \sum_{\substack{k \in \mathbb{N}^N \\ |k|_\infty < 2^j}} K(x, k/2^j) v(k/2^j) \right\|_\infty \to 0, \ j \to \infty.$$

It follows from (∗) that

$$\left| \langle \mu, u \rangle - 2^{-jN} \sum_{\substack{k \in \mathbb{N}^N \\ |k|_\infty < 2^j}} g(k/2^j) v(k/2^j) \right| \to 0, \ j \to \infty.$$

Since

$$\left| \int_\omega g(y) v(y) dy - 2^{-jN} \sum_{\substack{k \in \mathbb{N}^N \\ |k|_\infty < 2^j}} g(k/2^j) v(k/2^j) \right| \to 0, \ j \to \infty,$$

we conclude that $\langle \mu, u \rangle = \int_\omega g(y) v(y) dy.$  □

**Proof** (of Theorem 7.5.14.) Lemmas 7.5.15 and 7.5.16 imply the existence of $\alpha \in \mathbb{N}^N$ and of a finite measure $\mu$ on $\omega$ such that $f = \partial^\alpha \mu$ on $\mathcal{D}(\omega)$. By Lemma 7.5.17 there exists $g \in C_0(\omega)$ such that $\mu = \partial^{(2,\dots,2)} g$ on $\mathcal{D}(\omega)$.  □

## 7.6  Comments

The book by Maz'ya [51] is the main reference on functions of bounded variations and on capacity theory. The derivative of the function of unbounded variation in Sect. 7.3 is Denjoy–Perron integrable (since it is a derivative); see *Analyse, fondements techniques, évolution* by J. Mawhin [49].

## 7.7  Exercises for Chap. 7

1. Let $1 \leq p < N$. Then

$$\lambda p + N < 0 \Leftrightarrow (1+|x|^2)^{\lambda/2} \in W^{1,p}(\mathbb{R}^N),$$

$$(\lambda - 1)p + N < 0 \Leftrightarrow (1+|x|^2)^{\lambda/2} \in \mathcal{D}^{1,p}(\mathbb{R}^N).$$

2. What are the interior and the closure of $W^{1,1}(\Omega)$ in $BV(\Omega)$?
3. Let $u \in L^1_{\text{loc}}(\Omega)$. The following properties are equivalent:

   (a) $||Du||_\Omega < \infty$;
   (b) there exists $c > 0$ such that for every $\omega \subset\subset \Omega$ and every $y \in \mathbb{R}^N$ such that
   $|y| < d(\omega, \partial\Omega)$

$$||\tau_y u - u||_{L^1(\omega)} \le c|y|.$$

4. Relative variational capacity. Let $\Omega$ be an open bounded subset of $\mathbb{R}^N$ (or more generally, an open subset bounded in one direction). Let $1 \le p < \infty$ and let $K$ be a compact subset of $\Omega$. The capacity of degree $p$ of $K$ relative to $\Omega$ is defined by

$$\text{cap}_{p,\Omega}(K) = \inf\left\{\int_\Omega |\nabla u|^p dx : u \in W_K^{1,p}(\Omega)\right\},$$

where

$$W_K^{1,p}(\Omega) = \{u \in W_0^{1,p}(\Omega) : \text{there exists } \omega \text{ such that } K \subset \omega \subset\subset \Omega$$

$$\text{and } \chi_\omega \le u \text{ a.e. on } \Omega\}.$$

   Prove that the capacity of degree $p$ relative to $\Omega$ is a capacity on $\Omega$.
5. Verify that

$$\text{cap}_{p,\Omega}(K) = \inf\left\{\int_\Omega |\nabla u|^p dx : u \in \mathcal{D}_K(\Omega)\right\},$$

where

$$\mathcal{D}_K(\Omega) = \{u \in \mathcal{D}(\Omega) : \text{there exists } \omega \text{ such that } K \subset \omega \subset\subset \Omega \text{ and } \chi_\omega \le u\}.$$

6. (a) If $\text{cap}_{p,\Omega}(K) = 0$, then $m(K) = 0$. Hint: Use Poincaré's inequality.
   (b) If $p > N$ and if $\text{cap}_{p,\Omega}(K) = 0$, then $K = \phi$. Hint: Use Morrey inequalities.
7. Assume that $\text{cap}_{p,\Omega}(K) = 0$. Then for every $u \in \mathcal{D}(\Omega)$, there exists $(u_n) \subset \mathcal{D}(\Omega \setminus K)$ such that $|u_n| \le |u|$ and $u_n \to u$ in $W^{1,p}(\Omega)$.
8. Dupaigne–Ponce (2004). Assume that $\text{cap}_{1,\Omega}(K) = 0$. Then $W^{1,p}(\Omega \setminus K) = W^{1,p}(\Omega)$. Hint: Consider first the bounded functions in $W^{1,p}(\Omega \setminus K)$.
9. For every $u \in BV(\mathbb{R}^N)$, $||D|u|||_{\mathbb{R}^N} \le ||Du^+||_{\mathbb{R}^N} + ||Du^-||_{\mathbb{R}^N} = ||Du||_{\mathbb{R}^N}$.
   Hint: Consider a sequence $(u_n) \subset W^{1,1}(\mathbb{R}^N)$ such that $u_n \to u$ strictly in $BV(\mathbb{R}^N)$.

10. Let $u \in L^1(\Omega)$ and $f \in \mathcal{B}C^1(\Omega)$. Then $\|D(fu)\|_\Omega \leq \|f\|_\infty \|Du\|_\Omega + \|\nabla f\|_\infty \|u\|_{L^1(\Omega)}$.

11. Cheeger constant. Let $\Omega$ be an open subset of $\mathbb{R}^N$. Then for $1 \leq p < \infty$ and every $u \in W_0^{1,p}(\Omega)$,

$$\left(\frac{h(\Omega)}{p}\right)^p \int_\Omega |u|^p dx \leq \int_\Omega |\nabla u|^p dx.$$

12. Let $u \in W^{1,1}(\Omega)$. Then

$$\int_\Omega \sqrt{1+|\nabla u|^2} dx = \sup\left\{\int_\Omega (v_{N+1} + u\sum_{k=1}^N \partial_k v_k)dx : v \in \mathcal{D}(\Omega; \mathbb{R}^{N+1}), \|v\|_\infty \leq 1\right\}.$$

13. Support of a distribution. Let $f \in \mathcal{D}^*(\Omega)$ and $\omega \subset \Omega$. The restriction of $f$ to $\omega$ is zero if for all $u \in \mathcal{D}(\omega)$, $\langle f, u \rangle = 0$. The *support* of $f$, denoted by spt $f$, is the subset of $\Omega$ complementary to the largest open set in $\Omega$ on which the restriction of $f$ is zero. Prove that the support of $f$ is well defined.

14. Generalized divergence theorem. Let $A$ be a measurable subset of $\mathbb{R}^N$ such that $\|D\chi_A\|_{\mathbb{R}^N} < \infty$. Then spt $|D\chi_A| \subset \partial A$, and there exists $g: \mathbb{R}^N \to \mathbb{R}^N$ such that:

(a) $g$ is $|D\chi_A|$-measurable;
(b) $|g(x)| = 1$, $|D\chi_A|$-almost everywhere on $\mathbb{R}^N$;
(c) for all $v \in \mathcal{D}(\mathbb{R}^N; \mathbb{R}^N)$, $\int_A \operatorname{div} v \, dx = \int_{\mathbb{R}^N} v \cdot g d \, |D\chi_A|$.

15. Let $\Omega$ be an open subset of $\mathbb{R}^N$ of class $C^1$ with a bounded boundary or the product of $N$ open intervals. If $N \geq 2$ and if $1 \leq q \leq N/(N-1)$, then $BV(\Omega) \subset L^q(\Omega)$ and the canonical injection is continuous.

16. Let $\Omega$ be an open bounded subset of $\mathbb{R}^N$ of class $C^1$ or the product of $N$ bounded open intervals. If $N \geq 2$ and if $1 \leq q < N/(N-1)$, then $BV(\Omega) \subset L^q(\Omega)$ and the canonical injection is compact. Moreover Poincaré's inequality is valid: there exists $c = c(\Omega) > 0$ such that, for every $u \in BV(\Omega)$,

$$\left\| u - \int u \right\|_{L^1(\Omega)} \leq c \, \|Du\|_\Omega.$$

17. Let $1 < p < +\infty$ and $k \geq 1$. Define $M = \sum_{|\alpha| \leq k} 1$. The space $W^{-k,p'}(\Omega)$ is the space of distributions

$$g = \sum_{|\alpha| \leq k} (-1)^{|\alpha|} \partial^\alpha g_\alpha,$$

where $g_\alpha \in L^{p'}(\Omega; \mathbb{R}^M)$. Prove that $g \in W^{-k,p'}(\Omega)$ if and only if $g$ is the restriction to $\mathcal{D}(\Omega)$ of $f \in \left(W_0^{k,p}(\Omega)\right)^*$.

# Chapter 8
# Elliptic Problems

## 8.1 The Laplacian

The *Laplacian*, defined by

$$\Delta u = \operatorname{div} \nabla u = \frac{\partial^2 u}{\partial x_1^2} + \ldots + \frac{\partial^2 u}{\partial x_N^2},$$

is related to the mean of functions.

**Definition 8.1.1** Let $\Omega$ be an open subset of $\mathbb{R}^N$ and $u \in L^1_{\text{loc}}(\Omega)$. The mean of $u$ is defined on

$$D = \{(x, r) : x \in \Omega, 0 < r < d(x, \partial\Omega)\}$$

by

$$M(x, r) = V_N^{-1} \int_{B_N} u(x + ry)dy.$$

**Lemma 8.1.2** *Let $u \in C^2(\Omega)$. The mean of $u$ satisfies on $D$ the relation*

$$\lim_{r \downarrow 0} 2\frac{N+2}{r^2}[M(x, r) - u(x)] = \Delta u(x).$$

**Proof** Since we have uniformly for $|y| < 1$,

$$u(x + ry) = u(x) + r\nabla u(x) \cdot y + \frac{r^2}{2}D^2u(x)(y, y) + o(r^2),$$

© Springer Nature Switzerland AG 2022
M. Willem, *Functional Analysis*, Cornerstones,
https://doi.org/10.1007/978-3-031-09149-0_8

we obtain by symmetry

$$\int_{B_N} x_j dx = 0, \int_{B_N} x_j x_k dx = 0, \ j \neq k, \int_{B_N} x_j^2 dx = \frac{V_N}{N+2},$$

and

$$M(x,r) = u(x) + \frac{r^2}{2} \frac{1}{N+2} \Delta u(x) + o(r^2). \qquad \square$$

**Lemma 8.1.3** *Let* $u \in C^2(\Omega)$. *The following properties are equivalent:*

(a) $\Delta u \leq 0$;
(b) *for all* $(x,r) \in D$, $M(x,r) \leq u(x)$.

**Proof** By the preceding lemma, (a) follows from (b).

Assume that (a) is satisfied. Differentiating under the integral sign and using the divergence theorem, we obtain

$$\frac{\partial M}{\partial r}(x,r) = V_N^{-1} \int_{B_N} \nabla u(x+ry) \cdot y dy = r V_N^{-1} \int_{B_N} \Delta u(x+ry) \frac{1-|y|^2}{2} dy \leq 0.$$

We conclude that

$$M(x,r) \leq \lim_{r \downarrow 0} M(x,r) = u(x). \qquad \square$$

**Definition 8.1.4** Let $u \in L^1_{\text{loc}}(\Omega)$. The function $u$ is superharmonic if for every $v \in \mathcal{D}(\Omega)$ such that $v \geq 0$, $\int_{\Omega} u \Delta v dx \leq 0$.

The function $u$ is subharmonic if $-u$ is superharmonic.

The function $u$ is harmonic if for every $v \in \mathcal{D}(\Omega)$, $\int_{\Omega} u \Delta v dx = 0$.

We extend Lemma 8.1.3 to locally integrable functions.

**Theorem 8.1.5 (Mean-Value Inequality)** *Let* $u \in L^1_{\text{loc}}(\Omega)$. *The following properties are equivalent:*

(a) *u is superharmonic;*
(b) *for almost all* $x \in \Omega$ *and for all* $0 < r < d(x, \partial \Omega)$, $M(x,r) \leq u(x)$.

**Proof** Let $u_n = \rho_n * u$. Property (a) is equivalent to

(c) for every $n$, $\Delta u_n \leq 0$ on $\Omega_n$.

Property (b) is equivalent to

(d) for all $x \in \Omega_n$ and for all $0 < r < d(x, \partial \Omega_n)$, $V_N^{-1} \int_{B_N} u_n(x+ry) dy \leq u_n(x)$.

We conclude the proof using Lemma 8.1.3.

(a) $\Rightarrow$ (c). By Proposition 4.3.6, we have on $\Omega_n$ that

$$\Delta u_n(x) = \Delta \rho_n * u(x) = \int_\Omega \left(\Delta \rho_n(x-y)\right) u(y) dy \le 0.$$

(c) $\Rightarrow$ (a). It follows from the regularization theorem that for every $v \in \mathcal{D}(\Omega)$, $v \ge 0$,

$$\int_\Omega u \Delta v dx = \lim_{n\to\infty} \int_\Omega u_n \Delta v dx = \lim_{n\to\infty} \int_\Omega (\Delta u_n) v dx \le 0.$$

(b) $\Rightarrow$ (d). We have on $\Omega_n$ that

$$V_N^{-1} \int_{B_N} u_n(x+ry) dy = V_N^{-1} \int_{B(0,1/n)} dz \int_{B_N} \rho_n(z) u(x+ry-z) dy$$

$$\le \int_{B(0,1/n)} \rho_n(z) u(x-z) dz = u_n(x).$$

(d) $\Rightarrow$ (b). For $j \ge 1$, we define

$$\omega_j = \{x \in \Omega : d(x, \partial\Omega) > 1/j \text{ and } |x| < j\}.$$

Proposition 4.2.10 and the regularization theorem imply the existence of a subsequence $(u_{n_k})$ converging to $u$ in $L^1(\omega_j)$ and almost everywhere on $\omega_j$. Hence for almost all $x \in \omega_j$ and for all $0 < r < d(x, \partial\omega_j)$, $M(x,r) \le u(x)$. Since $\Omega = \bigcup_{j=1}^\infty \omega_j$, property (b) is satisfied. $\qquad\square$

**Theorem 8.1.6 (Maximum Principle)** *Let $\Omega$ be an open connected subset of $\mathbb{R}^N$ and $u \in L^1_{\text{loc}}(\Omega)$ a superharmonic function such that $u \ge 0$ almost everywhere on $\Omega$ and $u = 0$ on a subset of $\Omega$ with positive measure. Then $u = 0$ almost everywhere on $\Omega$.*

**Proof** Define

$$U_1 = \{x \in \Omega : \text{there exists } 0 < r < d(x, \partial\Omega) \text{ such that } M(x,r) = 0\}.$$

$$U_2 = \{x \in \Omega : \text{there exists } 0 < r < d(x, \partial\Omega) \text{ such that } M(x,r) > 0\}.$$

It is clear that $U_1$ and $U_2$ are open subsets of $\Omega$ such that $\Omega = U_1 \cup U_2$. By the preceding theorem, we obtain

$$U_2 = \{x \in \Omega : \text{for all } 0 < r < d(x, \partial\Omega), \, M(x, r) > 0\},$$

so that $U_1$ and $U_2$ are disjoint. If $\Omega = U_2$, then $u > 0$ almost everywhere on $\Omega$ by the preceding theorem. We conclude that $\Omega = U_1$ and $u = 0$ almost everywhere on $\Omega$. □

## 8.2   Eigenfunctions

> *En nous servant de quelques conceptions de l'analyse*
> *fonctionnelle nous représentons notre problème dans une forme*
> *nouvelle et démontrons que dans cette forme le problème admet*
> *toujours une solution unique.*
> *Si la solution cherchée existe dans le sens classique, alors notre*
> *solution se confond avec celle-ci.*
>
> S.L. Sobolev

Let $\Omega$ be a smooth bounded open subset of $\mathbb{R}^N$ with frontier $\Gamma$. An *eigenfunction* corresponding to the *eigenvalue* $\lambda$ is a nonzero solution of the problem

$$\begin{cases} -\Delta u = \lambda u & \text{in } \Omega, \\ u = 0 & \text{on } \Gamma. \end{cases} \qquad (\mathcal{P})$$

We will use the following *weak formulation* of problem $(\mathcal{P})$: find $u \in H_0^1(\Omega)$ such that for all $v \in H_0^1(\Omega)$,

$$\int_\Omega \nabla u \cdot \nabla v \, dx = \lambda \int_\Omega uv \, dx.$$

**Theorem 8.2.1** *There exist an unbounded sequence of eigenvalues of* $(\mathcal{P})$

$$0 < \lambda_1 \le \lambda_2 \le \cdots,$$

*and a sequence of corresponding eigenfunctions that is a Hilbert basis of* $H_0^1(\Omega)$.

***Proof*** On the space $H_0^1(\Omega)$, we define the inner product

$$a(u, v) = \int_\Omega \nabla u \cdot \nabla v \, dx$$

and the corresponding norm $\|u\|_a = \sqrt{a(u, u)}$.

For every $u \in H_0^1(\Omega)$, there exists one and only one $Au \in H_0^1(\Omega)$ such that for all $v \in H_0^1(\Omega)$,

$$a(Au, v) = \int_\Omega uv \, dx.$$

Hence problem $(\mathcal{P})$ is equivalent to

$$\lambda^{-1} u = Au.$$

Since $a(Au, u) = \int_\Omega u^2 dx$, the eigenvalues of $A$ are strictly positive. The operator $A$ is symmetric, since

$$a(Au, v) = \int_\Omega uv \, dx = a(u, Av).$$

It follows from the Cauchy–Schwarz and Poincaré inequalities that

$$||Au||_a^2 = \int_\Omega u \, Au \, dx \leq ||u||_{L^2(\Omega)} ||Au||_{L^2(\Omega)} \leq c||u||_{L^2(\Omega)} ||Au||_a.$$

Hence

$$||Au||_a \leq c||u||_{L^2(\Omega)}.$$

By the Rellich–Kondrachov theorem, the embedding $H_0^1(\Omega) \to L^2(\Omega)$ is compact, so that the operator $A$ is compact. We conclude using Theorem 3.4.8. □

**Proposition 8.2.2 (Poincaré's Principle)** *For every $n \geq 1$,*

$$\lambda_n = \min\left\{\int_\Omega |\nabla u|^2 dx : u \in H_0^1(\Omega), \int_\Omega u^2 dx = 1, \int_\Omega u e_1 dx = \ldots = \int_\Omega u e_{n-1} dx = 0\right\}.$$

***Proof*** We deduce from Theorem 3.4.7 that

$$\lambda_n^{-1} = \max\left\{\frac{a(Au, u)}{a(u, u)} : u \in H_0^1(\Omega), u \neq 0, a(u, e_1) = \ldots = a(u, e_{n-1}) = 0\right\}.$$

Since $e_k$ is an eigenfunction,

$$a(u, e_k) = 0 \iff \int_\Omega u e_k dx = 0.$$

Hence we obtain

$$\lambda_n^{-1} = \max\left\{\frac{\int_\Omega u^2 dx}{\int_\Omega |\nabla u|^2 dx} : u \in H_0^1(\Omega), u \neq 0, \int_\Omega u e_1 dx = \ldots = \int_\Omega u e_{n-1} dx = 0\right\},$$

or

$$\lambda_n = \min\left\{\frac{\int_\Omega |\nabla u|^2 dx}{\int_\Omega u^2 dx} : u \in H_0^1(\Omega), u \neq 0, \int_\Omega u e_1 dx = \ldots = \int_\Omega u e_{n-1} dx = 0\right\}. \ \square$$

**Proposition 8.2.3** *Let $u \in H_0^1(\Omega)$ be such that $||u||_2 = 1$ and $||\nabla u||_2^2 = \lambda_1$. Then $u$ is an eigenfunction corresponding to the eigenvalue $\lambda_1$.*

***Proof*** Let $v \in H_0^1(\Omega)$. The function

$$g(\varepsilon) = ||\nabla(u + \varepsilon v)||_2^2 - \lambda_1 ||u + \varepsilon v||_2^2$$

reaches its minimum at $\varepsilon = 0$. Hence $g'(0) = 0$ and

$$\int_\Omega \nabla u \cdot \nabla v \, dx - \lambda_1 \int_\Omega uv \, dx = 0. \qquad\qquad \square$$

**Proposition 8.2.4** *Let $\Omega$ be a smooth bounded open connected subset of $\mathbb{R}^N$. Then the eigenvalue $\lambda_1$ of $(\mathcal{P})$ is simple, and $e_1$ is almost everywhere strictly positive on $\Omega$.*

***Proof*** Let $u$ be an eigenfunction corresponding to $\lambda_1$ and such that $||u||_2 = 1$. By Corollary 6.1.14, $v = |u| \in H_0^1(\Omega)$ and $||\nabla v||_2^2 = ||\nabla u||_2^2 = \lambda_1$. Since $||v||_2 = ||u||_2 = 1$, the preceding proposition implies that $v$ is an eigenfunction corresponding to $\lambda_1$. Assume that $u^+ \neq 0$. Then $u^+$ is an eigenfunction corresponding to $\lambda_1$, and by the maximum principle, $u^+ > 0$ almost everywhere on $\Omega$. Hence $u = u^+$. Similarly, if $u^- \neq 0$, then $-u = u^- > 0$ almost everywhere on $\Omega$. We can assume that $e_1 > 0$ almost everywhere on $\Omega$. If $e_2$ corresponds to $\lambda_1$, then $e_2$ is either positive or negative, and $\int_\Omega e_1 e_2 dx = 0$. This is a contradiction. $\qquad \square$

*Example* Let $\Omega = \,]0, \pi[$. Then $(\mathcal{P})$ becomes

$$\begin{cases} -u'' = \lambda u & \text{in } ]0, \pi[, \\ u(0) = u(\pi) = 0. \end{cases}$$

Sobolev's embedding theorem and the du Bois–Reymond lemma imply that $u \in C^2(]0, \pi[) \cap C([0, \pi])$. Hence $\lambda_n = n^2$ and $e_n = \sqrt{\frac{2}{\pi}} \frac{\sin nx}{n}$. The sequence $(e_n)$ is a

Hilbert basis on $H_0^1(]0, \pi[)$ with scalar product $\int_0^\pi u'v'\,dx$, and the sequence $(ne_n)$ is a Hilbert basis of $L^2(]0, \pi[)$ with scalar product $\int_0^\pi uv\,dx$.

**Definition 8.2.5** Let $G$ be a subgroup of the orthogonal group $\mathbf{O}(N)$. The open subset $\Omega$ of $\mathbb{R}^N$ is $G$-invariant if for every $g \in G$ and every $x \in \Omega$, $g^{-1}x \in \Omega$. Let $\Omega$ be $G$-invariant. The action of $G$ on $H_0^1(\Omega)$ is defined by $gu(x) = u(g^{-1}x)$. The space of fixed points of $G$ is defined by

$$\mathrm{Fix}(G) = \{u \in H_0^1(\Omega) : \text{for every } g \in G, gu = u\}.$$

A function $J : H_0^1(\Omega) \to \mathbb{R}$ is $G$-invariant if for every $g \in G$, $J \circ g = J$.

**Proposition 8.2.6** Let $\Omega$ be a $G$-invariant open subset of $\mathbb{R}^N$ satisfying the assumptions of Proposition 8.2.4. Then $e_1 \in \mathrm{Fix}(G)$.

**Proof** By a direct computation, we obtain, for all $g \in G$,

$$||ge_1||_2 = ||e_1||_2 = 1, ||\nabla ge_1||_2^2 = ||\nabla e_1||_2^2 = \lambda_1.$$

Propositions 8.2.3 and 8.2.4 imply the existence of a scalar $\lambda(g)$ such that

$$e_1(g^{-1}x) = \lambda(g)e_1(x).$$

Integrating on $\Omega$, we obtain $\lambda(g) = 1$. But then $ge_1 = e_1$. Since $g \in G$ is arbitrary, $e_1 \in \mathrm{Fix}(G)$.                                                                □

*Example (Symmetry of the First Eigenfunction)* For a ball or an annulus

$$\Omega = \{x \in \mathbb{R}^N : r < |x| < R\},$$

we choose $G = \mathbf{O}(N)$. Hence $e_1$ is a radial function.

We define $v(|x|) = u(x)$. By a simple computation, we have

$$\frac{\partial^2}{\partial x_k^2}u(x) = v''(|x|)\frac{x_k^2}{|x|^2} + v'(|x|)\left(\frac{1}{|x|} - \frac{x_k^2}{|x|^3}\right).$$

Hence we obtain

$$\Delta u = v'' + (N - 1)v'/|x|.$$

Let $\Omega = B(0, 1) \subset \mathbb{R}^3$. The first eigenfunction, $u(x) = v(|x|)$, is a solution of

$$-v'' - 2v'/r = \lambda v.$$

The function $w = rv$ satisfies

$$-w'' = \lambda w,$$

so that

$$w(r) = a \sin(\sqrt{\lambda} r - b)$$

and

$$v(r) = a \frac{\sin(\sqrt{\lambda} r - b)}{r}.$$

Since $u \in H_0^1(\Omega) \subset L^6(\Omega)$, $b = 0$ and $\lambda = \pi^2$. Finally, we obtain

$$u(x) = a \frac{\sin(\pi |x|)}{|x|}.$$

It follows from Poincaré's principle that

$$\pi^2 = \min \left\{ ||\nabla u||^2_{L^2(\Omega)}/||u||^2_{L^2(\Omega)} : u \in H_0^1(\Omega) \setminus \{0\} \right\}.$$

Let us characterize the eigenvalues without using the eigenfunctions.

**Theorem 8.2.7 (Max-inf Principle)**   *For every $n \geq 1$,*

$$\lambda_n = \max_{V \in \mathcal{V}_{n-1}} \inf_{\substack{u \in V^\perp \\ ||u||_{L^2} = 1}} \int_\Omega |\nabla u|^2 dx,$$

*where $\mathcal{V}_{n-1}$ denotes the family of all $(n-1)$-dimensional subspaces of $H_0^1(\Omega)$.*

**Proof** Let us denote by $\Lambda_n$ the second member of the preceding equality. It follows from Poincaré's principle that $\lambda_n \leq \Lambda_n$.

Let $V \in \mathcal{V}_{n-1}$. Since the codimension of $V^\perp$ is equal to $n - 1$, there exists $x \in \mathbb{R}^N \setminus \{0\}$ such that $u = \sum_{j=1}^n x_j e_j \in V^\perp$. Since

$$\int_\Omega |\nabla u|^2 dx = \sum_{j=1}^n \lambda_j x_j^2 \int_\Omega e_j^2 dx \leq \lambda_n \int_\Omega u^2 dx,$$

we obtain

$$\inf_{\substack{u \in V^\perp \\ \|u\|_{L^2} = 1}} \int_\Omega |\nabla u|^2 dx \leq \lambda_n.$$

Since $V \in \mathcal{V}_{n-1}$ is arbitrary, we conclude that $\Lambda_n \leq \lambda_n$.                                    $\square$

## 8.3   Symmetrization

> *La considération systématique des ensembles $E[a \leq f < b]$*
> *m'a été pratiquement utile parce qu'elle m'a toujours forcé à*
> *grouper les conditions donnant des effets voisins.*
>
> *Henri Lebesgue*

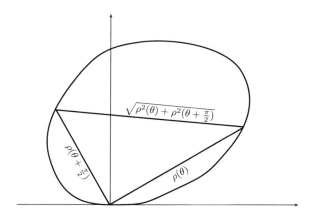

**Fig. 8.1** Isodiametric inequality

According to the *isodiametric inequality* in $\mathbb{R}^2$, among all domains with a fixed diameter, the disk has the largest area. A simple proof was given by J.E. Littlewood in 1953 in *A Mathematician's Miscellany*. We can assume that the domain $\Omega$ is convex and that the horizontal axis is tangent to $\Omega$ at the origin. We obtain

$$A = \frac{1}{2} \int_0^{\frac{\pi}{2}} \rho^2(\theta) + \rho^2 \left( \theta + \frac{\pi}{2} \right) d\theta \leq \pi (d/2)^2.$$

We will prove the *isoperimetric inequality* in $\mathbb{R}^N$ using *Schwarz's symmetrization*.

In this section, we consider Lebesgue's measure on $\mathbb{R}^N$. We define

$$\mathcal{K}_+(\mathbb{R}^N) = \{u \in \mathcal{K}(\mathbb{R}^N) : \text{for all } x \in \mathbb{R}^N, u(x) \geq 0\},$$
$$L^p_+(\mathbb{R}^N) = \{u \in L^p(\mathbb{R}^N) : \text{for almost all } u(x) \geq 0\},$$
$$W^{1,p}_+(\mathbb{R}^N) = W^{1,p}(\mathbb{R}^N) \cap L^p_+(\mathbb{R}^N),$$
$$BV_+(\mathbb{R}^N) = BV(\mathbb{R}^N) \cap L^1_+(\mathbb{R}^N).$$

**Definition 8.3.1** Schwarz's symmetrization of a measurable subset $A$ of $\mathbb{R}^N$ is defined by $A^* = \{x \in \mathbb{R}^N : |x|^N V_N < m(A)\}$. An admissible function $u : \mathbb{R}^N \to [0, +\infty]$ is a measurable function such that for all $t > 0$, $m_u(t) = m(\{u > t\}) < \infty$. Schwarz's symmetrization of an admissible function $u$ is defined on $\mathbb{R}^N$ by

$$u^*(x) = \sup\{t \in \mathbb{R} : x \in \{u > t\}^*\}.$$

The following properties are clear:

(a)  $\chi_{A^*} = \chi_A^*$;
(b)  $m(A^* \setminus B^*) \leq m(A \setminus B)$;
(c)  $u^*$ is radially decreasing, $|x| \leq |y| \Rightarrow u^*(x) \geq u^*(y)$;
(d)  $u \leq v \Rightarrow u^* \leq v^*$.

**Lemma 8.3.2** *Let $(A_n)$ be an increasing sequence of measurable sets. Then*

$$\bigcup_{n=1}^{\infty} A_n^* = \left(\bigcup_{n=1}^{\infty} A_n\right)^*.$$

**Proof** By definition, $A_n^* = B(0, r_n)$, $\left(\bigcup_{n=1}^{\infty} A_n\right)^* = B(0, r)$, where $r_n^N V_N = m(A_n)$, $r^N V_N = m\left(\bigcup_{n=1}^{\infty} A_n\right)$. It suffices to observe that by Proposition 2.2.26,

$$m\left(\bigcup_{n=1}^{\infty} A_n\right) = \lim_{n \to \infty} m(A_n). \qquad \square$$

**Theorem 8.3.3** *Let $u$ be an admissible function. Then $u^*$ is lower semicontinuous, and for all $t > 0$, $\{u > t\}^* = \{u^* > t\}$ and $m_u(t) = m_{u^*}(t)$.*

**Proof** Let $t > 0$. Using the preceding lemma, we obtain

$$\{u > t\}^* = \left(\bigcup_{s > t}\{u > s\}\right)^* = \bigcup_{s > t}\{u > s\}^* \subset \{u^* > t\} \subset \{u > t\}^*.$$

In particular, $\{u^* > t\}$ is open and $m\{u > t\} = m\{u^* > t\}$. $\qquad \square$

**Proposition 8.3.4** *Let $1 \leq p < \infty$ and $u, v \in L^p_+(\mathbb{R}^N)$. Then $u^*, v^* \in L^p_+(\mathbb{R}^N)$ and*

$$||u^*||_p = ||u||_p, ||u^* - v^*||_p \leq ||u - v||_p.$$

**Proof** Using Cavalieri's principle and the preceding theorem, we obtain

$$||u^*||_p^p = \int_0^\infty m_{(u^*)^p}(t)dt = \int_0^\infty m_{u^p}(t)dt = ||u||_p^p.$$

Assume that $p \geq 2$, and define $g(t) = |t|^p$, so that $g$ is convex, even, of class $C^2$, and $g(0) = g'(0) = 0$. For $a < b$, the fundamental theorem of calculus implies that

$$g(b - a) = \int_a^b ds \int_s^b g''(t - s)dt.$$

Hence we have that

$$g(u - v) = \int_0^\infty ds \int_s^\infty g''(t - s) \left[ \chi_{\{u>t\}}(1 - \chi_{\{v>s\}}) + \chi_{\{v>t\}}(1 - \chi_{\{u>s\}}) \right] dt.$$

Integrating on $\mathbb{R}^N$ and using Fubini's theorem, we find that

$$\int_{\mathbb{R}^N} g(u-v)dx = \int_0^\infty ds \int_s^\infty g''(t-s)[m(\{u > t\}\backslash\{v > s\})+m(\{v > t\}\backslash\{u > s\})]dt.$$

Finally, we obtain

$$\int_{\mathbb{R}^N} g(u^* - v^*)dx \leq \int_{\mathbb{R}^N} g(u - v)dx.$$

If $1 \leq p < 2$, it suffices to approximate $|t|^p$ by $g_\varepsilon(t) = (t^2 + \varepsilon^2)^{p/2} - \varepsilon^p, \varepsilon > 0$.   □

Approximating *Schwarz's symmetrizations* by *polarizations*, we will prove that if $u \in W^{1,p}_+(\mathbb{R}^N)$, then $u^* \in W^{1,p}_+(\mathbb{R}^N)$ and $||\nabla u^*||_p \leq ||\nabla u||_p$.

**Definition 8.3.5** Let $\sigma_H$ be the reflection with respect to the frontier of a closed affine half-space $H$ of $\mathbb{R}^N$. The polarization (with respect to $H$) of a function $u : \mathbb{R}^N \rightarrow \mathbb{R}$ is defined by

$$u^H(x) = \max\{u(x), u(\sigma_H(x))\}, \quad x \in H,$$
$$= \min\{u(x), u(\sigma_H(x))\}, \quad x \in \mathbb{R}^N \setminus H.$$

The polarization $A^H$ of $A \subset \mathbb{R}^N$ is defined by $\chi_{A^H} = \chi_A^H$. We denote by $\mathcal{H}$ the family of all closed affine half-spaces of $\mathbb{R}^N$ containing 0.

Let us recall that a closed affine half-space of $\mathbb{R}^N$ is defined by

$$H = \{x \in \mathbb{R}^N : a \cdot x \le b\},$$

where $a \in \mathbb{S}^{N-1}$ and $b \in \mathbb{R}$. It is clear that

$$\sigma_H(x) = x + 2(b - a \cdot x)a.$$

The following properties are easy to prove:

(a) if $A$ is a measurable subset of $\mathbb{R}^N$, then $m(A^H) = m(A)$;
(b) $\{u^H > t\} = \{u > t\}^H$;
(c) if $u$ is admissible, $(u^H)^* = u^*$;
(d) if moreover, $H \in \mathcal{H}$, $(u^*)^H = u^*$.

**Lemma 8.3.6** *Let $f : \mathbb{R} \to \mathbb{R}$ be convex and $a \le b, c \le d$. Then*

$$f(b - d) + f(a - c) \le f(a - d) + f(b - c).$$

***Proof*** Define $x = b - d$, $y = b - a$, and $z = d - c$. By convexity, we have

$$f(x) - f(x - y) \le f(x + z) - f(x + z - y). \qquad \square$$

**Proposition 8.3.7** *Let $1 \le p < \infty$ and $u, v \in L^p(\mathbb{R}^N)$. Then $u^H, v^H \in L^p(\mathbb{R}^N)$, and*

$$\|u^H\|_p = \|u\|_p, \quad \|u^H - v^H\|_p \le \|u - v\|_p.$$

***Proof*** Observe that

$$\int_{\mathbb{R}^N} |u(x)|^p dx = \int_H |u(x)|^p + |u(\sigma_H(x))|^p dx$$

$$= \int_H |u^H(x)|^p + |u^H(\sigma_H(x))|^p dx = \int_{\mathbb{R}^N} |u^H(x)|^p dx.$$

Using the preceding lemma, it is easy to verify that for all $x \in H$,

$$|u^H(x) - v^H(x)|^p + |u^H(\sigma_H(x)) - v^H(\sigma_H(x))|^p$$
$$\le |u(x) - v(x)|^p + |u(\sigma_H(x)) - v(\sigma_H(x))|^p.$$

It suffices then to integrate over $H$. $\qquad \square$

**Lemma 8.3.8** *Let $u : \mathbb{R}^N \to \mathbb{R}$ be a uniformly continuous function. Then the function $u^H : \mathbb{R}^N \to \mathbb{R}$ is uniformly continuous, and for all $\delta > 0$, $\omega_{u^H}(\delta) \le \omega_u(\delta)$.*

**Proof** Let $\delta > 0$ and $x, y \in \mathbb{R}^N$ be such that $|x - y| \le \delta$. If $x, y \in H$ or if $x, y \in \mathbb{R}^N \setminus H$, we have

$$|\sigma_H(x) - \sigma_H(y)| = |x - y| \le \delta$$

and

$$|u^H(x) - u^H(y)| \le \max(|u(x) - u(y)|, |u(\sigma_H(x)) - u(\sigma_H(y))|) \le \omega_u(\delta).$$

If $x \in H$ and $y \in \mathbb{R}^N \setminus H$, we have

$$|x - \sigma_H(y)| = |\sigma_H(x) - y| \le |\sigma_H(x) - \sigma_H(y)| = |x - y| \le \delta$$

and

$$|u^H(x) - u^H(y)| \le \max(|u(x) - u(\sigma_H(y))|, |u(\sigma_H(x)) - u(y)|,$$
$$|u(\sigma_H(x)) - u(\sigma_H(y))|, |u(x) - u(y)|) \le \omega_u(\delta).$$

We conclude that

$$\omega_{u^H}(\delta) = \sup_{|x-y| \le \delta} |u^H(x) - u^H(y)| \le \omega_u(\delta). \qquad \square$$

**Lemma 8.3.9** *Let $1 \le p < \infty$, $u \in L^p(\mathbb{R}^N)$, and $H \in \mathcal{H}$. Define $g(x) = e^{-|x|^2}$. Then*

$$\int_{\mathbb{R}^N} ug \, dx \le \int_{\mathbb{R}^N} u^H g \, dx. \qquad (*)$$

*If, moreover, $0 \in \overset{o}{H}$ and*

$$\int_{\mathbb{R}^N} ug \, dx = \int_{\mathbb{R}^N} u^H g \, dx, \qquad (**)$$

*then $u^H = u$.*

**Proof** For all $x \in H$, we have

$$u(x)g(x) + u(\sigma_H(x))g(\sigma_H(x)) \le u^H(x)g(x) + u^H(\sigma_H(x))g(\sigma_H(x)).$$

It suffices then to integrate over $H$ to prove $(*)$.

If $(**)$ holds, we obtain, almost everywhere on $H$,

$$u(x)g(x) + u(\sigma_H(x))g(\sigma_H(x)) = u^H(x)g(x) + u^H(\sigma_H(x))g(\sigma_H(x)).$$

If $0 \in \overset{o}{H}$, then $g(\sigma_H(x)) < g(x)$ for all $x \in \overset{o}{H}$, so that

$$u(x) = u^H(x), u(\sigma_H(x)) = u^H(\sigma_H(x)). \qquad \square$$

**Lemma 8.3.10** *Let* $u \in L^p(\mathbb{R}^N) \cap C(\mathbb{R}^N)(1 \leq p < \infty)$ *be such that, for all* $H \in \mathcal{H}, u^H = u$. *Then* $u \geq 0$ *and* $u = u^*$.

***Proof*** Let $x, y \in \mathbb{R}^N$ be such that $x \neq y$ and $|x| \leq |y|$. There exists $H \in \mathcal{H}$ such that $x \in H$ and $y = \sigma_H(x)$. By assumption, we have

$$u(y) = u^H(y) \leq u^H(x) = u(x).$$

Hence

$$|x| \leq |y| \Rightarrow u(y) \leq u(x).$$

We conclude that there exists a (continuous) decreasing function $v : [0, +\infty[ \to \mathbb{R}$ such that $u(x) = v(|x|)$. Since $u \in L^p(\mathbb{R}^N)$, it is clear that

$$\lim_{r \to +\infty} v(r) = 0.$$

Hence $u \geq 0$ and for all $t > 0$, $\{u > t\} = \{u^* > t\}$, so that $u = u^*$. $\qquad \square$

Consider a sequence of closed affine half-spaces

$$H_n = \{x \in \mathbb{R}^N : a_n \cdot x \leq b_n\}$$

such that $((a_n, b_n))$ is dense in $\mathbb{S}^{N-1} \times ]0, +\infty[$.

The following result is due to J. Van Schaftingen.

**Theorem 8.3.11** *Let* $1 \leq p < \infty$ *and* $u \in L^p_+(\mathbb{R}^N)$. *Define*

$$u_0 = u,$$
$$u_{n+1} = u_n^{H_1 \ldots H_{n+1}}.$$

*Then the sequence* $(u_n)$ *converges to* $u^*$ *in* $L^p(\mathbb{R}^N)$.

***Proof*** Assume that $u \in \mathcal{K}_+(\mathbb{R}^N)$. There exists $r > 0$ such that spt $u \subset B[0, r]$. Hence for all $n$,

$$\text{spt } u_n \subset B[0, r].$$

The sequence $(u_n)$ is precompact in $C(B[0, r])$ by Ascoli's theorem, since

(a) for every $n$, $||u_n||_\infty = ||u||_\infty$;
(b) for every $\varepsilon > 0$, there exists $\delta > 0$, such that for every $n$, $\omega_{u_n}(\delta) \leq \omega_u(\delta) \leq \varepsilon$.

Assume that $(u_{n_k})$ converges uniformly to $v$. Observe that

$$\text{spt } v \subset B[0, r].$$

We shall prove that $v = u^*$. Since by Proposition 8.3.4,

$$||u^* - v^*||_1 = ||u^*_{n_k} - v^*||_1 \leq ||u_{n_k} - v||_1 \to 0, \quad k \to \infty,$$

it suffices to prove that $v = v^*$.
Let $m \geq 1$. For every $n_k \geq m$, we have

$$u_{n_{k+1}} = u_{n_k}^{H_1 \ldots H_m \ldots H_{n_{k+1}}}.$$

Lemma 8.3.9 implies that

$$\int_{\mathbb{R}^N} u_{n_k}^{H_1 \ldots H_m} g \, dx \leq \int_{\mathbb{R}^N} u_{n_{k+1}} g \, dx.$$

It follows from Proposition 8.3.7 that

$$\int_{\mathbb{R}^N} v^{H_1 \ldots H_m} g \, dx \leq \int_{\mathbb{R}^N} vg \, dx.$$

By Lemma 8.3.9, $v^{H_1} = v$, and by induction, $v^{H_m} = v$.
Let $a \in \mathbb{S}^{N-1}$, $b \geq 0$, and $H = \{x \in \mathbb{R}^N : a \cdot x \leq b\}$. There exists a sequence $(n_k)$ such that $(a_{n_k}, b_{n_k}) \to (a, b)$. We deduce from Lebesgue's dominated convergence theorem that

$$||v^H - v||_1 = ||v^H - v^{H_{n_k}}||_1 \to 0, \quad k \to \infty.$$

Hence for all $H \in \mathcal{H}$, $v = v^H$. Lemma 8.3.10 ensures that $v = v^*$.
Let $u \in L^p_+(\mathbb{R}^N)$ and $\varepsilon > 0$. The density theorem implies the existence of $w \in \mathcal{K}_+(\mathbb{R}^N)$ such that $||u - w||_p \leq \varepsilon$. By the preceding step, the sequence

$$w_0 = w,$$
$$w_{n+1} = w_n^{H_1 \ldots H_{n+1}},$$

converges to $w^*$ in $L^p(\mathbb{R}^N)$. Hence there exists $m$ such that for $n \geq m$, $||w_n - w^*||_p \leq \varepsilon$. It follows from Propositions 8.3.4 and 8.3.7 that for $n \geq m$,

$$||u_n - u^*||_p \leq ||u_n - w_n||_p + ||w_n - w^*||_p + ||w^* - u^*||_p \leq 2||u - w||_p + \varepsilon \leq 3\varepsilon.$$

Since $\varepsilon > 0$ is arbitrary, the proof is complete.                                                                    □

**Proposition 8.3.12** Let $1 \leq p < \infty$ and $u \in W^{1,p}(\mathbb{R}^N)$. Then $u^H \in W^{1,p}(\mathbb{R}^N)$ and $||\nabla u^H||_p = ||\nabla u||_p$.

**Proof** Define $v = u \circ \sigma_H$. Observe that

$$u^H = \frac{1}{2}(u + v) + \frac{1}{2}|u - v|, \quad \text{on } H,$$

$$= \frac{1}{2}(u + v) - \frac{1}{2}|u - v|, \quad \text{on } \mathbb{R}^N \setminus H.$$

Since the trace of $|u - v|$ is equal to 0 on $\partial H$, $u^H \in W^{1,p}(\mathbb{R}^N)$. Let $x \in H$. Corollary 6.1.14 implies that for $u(x) \geq v(x)$,

$$\nabla u^H(x) = \nabla u(x), \nabla u^H(\sigma_H(x)) = \nabla u(\sigma_H(x)),$$

and for $u(x) < v(x)$,

$$\nabla u^H(x) = \nabla v(x), \nabla u^H(\sigma_H(x)) = \nabla v(\sigma_H(x)).$$

We conclude that on $H$,

$$|\nabla u^H(x)|^p + |\nabla u^H(\sigma_H(x))|^p = |\nabla u(x)|^p + |\nabla u(\sigma_H(x))|^p. \qquad □$$

**Proposition 8.3.13** Let $u \in BV(\mathbb{R}^N)$. Then $u^H \in BV(\mathbb{R}^N)$ and $||Du^H|| \leq ||Du||$.

**Proof** Let $u_n = \rho_n * u$. Propositions 4.3.14 and 8.3.7 imply that $u_n \to u$ and $u_n^H \to u^H$ in $L^1(\mathbb{R}^N)$. Theorem 7.3.3 and Proposition 8.3.12 ensure that

$$||Du_n^H|| = ||\nabla u_n^H||_1 = ||\nabla u_n||_1.$$

We conclude by Theorem 7.3.2 and Lemma 7.3.6 that

$$||Du^H|| \leq \lim ||Du_n^H|| = \lim ||\nabla u_n||_1 = ||Du||. \qquad □$$

**Theorem 8.3.14 (Pólya–Szegő Inequality)** Let $1 < p < \infty$ and $u \in W_+^{1,p}(\mathbb{R}^N)$. Then $u^* \in W_+^{1,p}(\mathbb{R}^N)$ and $||\nabla u^*||_p \leq ||\nabla u||_p$.

**Proof** The sequence $(u_n)$ given by Theorem 8.3.11 converges to $u^*$ in $L^p(\mathbb{R}^N)$. By Proposition 8.3.12, for every $n$, $||\nabla u_n||_p = ||\nabla u||_p$. It follows from Theorem 6.1.7

that

$$||\nabla u^*||_p \leq \lim ||\nabla u_n||_p = ||\nabla u||_p.$$ □

**Theorem 8.3.15 (Hilden's Inequality, 1976)** *Let* $u \in BV_+(\mathbb{R}^N)$. *Then* $u^* \in BV_+(\mathbb{R}^N)$ *and* $||Du^*|| \leq ||Du||$.

***Proof*** The sequence $(u_n)$ given by Theorem 8.3.11 converges to $u^*$ in $L^{1^*}(\mathbb{R}^N)$. By Proposition 8.3.13, for every $n$,

$$||Du_{n+1}|| \leq ||Du_n|| \leq ||Du||.$$

It follows from Theorem 7.3.2 that

$$||Du^*|| \leq \lim ||Du_n|| \leq ||Du||.$$ □

**Theorem 8.3.16 (De Giorgi's Isoperimetric Inequality)** *Let* $N \geq 2$, *and let* $A$ *be a measurable subset of* $\mathbb{R}^N$ *with finite measure. Then*

$$N V_N^{1/N} (m(A))^{1-1/N} \leq p(A).$$

***Proof*** If $p(A) = +\infty$, the inequality is clear. If this is not the case, then $\chi_A \in BV_+(\mathbb{R}^N)$. By definition of Schwarz's symmetrization,

$$A^* = B(0, r), \quad V_N r^N = m(A).$$

Theorems 7.4.1 and 8.3.15 imply that

$$N V_N r^{N-1} = p(A^*) = ||D\chi_{A^*}||_{\mathbb{R}^N} = ||D\chi_A^*||_{\mathbb{R}^N} \leq ||D\chi_A||_{\mathbb{R}^N} = p(A).$$

It is easy to conclude the proof. □

Using scaling invariance, we obtain the following version of the isoperimetric inequality.

**Corollary 8.3.17** *Let* $A$ *be a measurable subset of* $\mathbb{R}^N$ *with finite measure, and let* $B$ *be an open ball of* $\mathbb{R}^N$. *Then*

$$p(B)/m(B)^{1-1/N} \leq p(A)/m(A)^{1-1/N}.$$

The constant $N V_N^{1/N}$, corresponding to the characteristic function of a ball, is the optimal constant for the Gagliardo–Nirenberg inequality.

**Theorem 8.3.18** *Let* $N \geq 2$ *and* $u \in L^{N/(N-1)}$ *such that* $||Du|| < +\infty$. *Then*

$$N V_N^{1/N} ||u||_{N/(N-1)} \leq ||Du||.$$

*Proof*

(a)  Let $p = N/(N-1)$, $v \in L^p(\mathbb{R}^N)$, $v \geq 0$, and $g \in L^{p'}(\mathbb{R}^N)$. If $||g||_{p'} = 1$, we deduce from Fubini's theorem and Hölder's inequality that

$$\int_{\mathbb{R}^N} gv dx = \int_{\mathbb{R}^N} dx \int_0^\infty g\chi_{v>t} dt = \int_0^\infty dt \int_{\mathbb{R}^N} g\chi_{v>t} dx \leq \int_0^\infty m(\{v > t\})^{1/p} dt.$$

Hence we obtain

$$||v||_p = \max_{||g||_{p'}=1} \int_{\mathbb{R}^N} gv dx \leq \int_0^\infty m(\{v > t\})^{1/p} dt. \qquad (*)$$

(b)  Let $u \in \mathcal{D}(\Omega)$. Using inequality $(*)$, the Morse–Sard theorem (Theorem 9.3.1), the coarea formula (Theorem 9.3.3), and the isoperimetric inequality, we obtain

$$N V_N^{1/N} ||u||_p \leq N V_N^{1/N} [||u^+||_p + ||u^-||_p]$$

$$\leq N V_N^{1/N} \left[ \int_0^\infty m(\{u > t\})^{1/p} dt + \int_{-\infty}^0 m(\{u < t\})^{1/p} dt \right]$$

$$\leq \int_0^\infty dt \int_{u=t} d\gamma + \int_{-\infty}^0 dt \int_{u=t} d\gamma = \int_{\mathbb{R}^N} |\nabla u| dx.$$

(c)  By density, we obtain, for every $u \in \mathcal{D}^{1,1}(\mathbb{R}^N)$,

$$N V_N^{1/N} ||u||_p \leq ||\nabla u||_1.$$

We conclude using Proposition 4.3.14 and Lemma 7.3.6.                           □

**Definition 8.3.19**  Let $\Omega$ be an open subset of $\mathbb{R}^N$. We define

$$\lambda_1(\Omega) = \inf \left\{ ||\nabla u||_2^2 / ||u||_2^2 : u \in W_0^{1,2}(\Omega) \setminus \{0\} \right\}.$$

**Theorem 8.3.20 (Faber–Krahn Inequality)**  *Let* $\Omega$ *be an open subset of* $\mathbb{R}^N$ *with finite measure. Then* $\lambda_1(\Omega^*) \leq \lambda_1(\Omega)$.

**Proof** Define $Q(u) = ||\nabla u||_2^2 / ||u||_2^2$. Let $u \in W_0^{1,2}(\Omega) \setminus \{0\}$ and $v = |u|$. By Corollary 6.1.14, $Q(v) = Q(u)$. Proposition 8.3.4 and the Pólya–Szegő inequality imply that $Q(v^*) \leq Q(v)$. It is easy to verify that $v^* \in W_0^{1,2}(\Omega^*) \setminus \{0\}$. Hence we obtain

$$\lambda_1(\Omega^*) \leq Q(v^*) \leq Q(v) = Q(u).$$

Since $u \in W_0^{1,2}(\Omega) \setminus \{0\}$ is arbitrary, it is easy to conclude the proof. $\square$

Using scaling invariance, we obtain the following version of the Faber–Krahn inequality.

**Corollary 8.3.21** *Let $\Omega$ be an open subset of $\mathbb{R}^N$, and let $B$ be an open ball of $\mathbb{R}^N$. Then*

$$\lambda_1(B)m(B)^{2/N} \leq \lambda_1(\Omega)m(\Omega)^{2/N}.$$

*Remark* Equality in the isoperimetric inequality or in the Faber–Krahn inequality is achieved only when the corresponding domain is a ball.

## 8.4 Elementary Solutions

There exists no locally integrable function corresponding to the *Dirac measure*.

**Definition 8.4.1** The Dirac measure is defined on $\mathcal{K}(\mathbb{R}^N)$ by

$$\langle \delta, u \rangle = u(0).$$

**Definition 8.4.2** The elementary solutions of the Laplacian are defined on $\mathbb{R}^N \setminus \{0\}$ by

$$E_N(x) = \frac{1}{2\pi} \log \frac{1}{|x|}, \qquad N = 2,$$

$$E_N(x) = \frac{1}{(N-2)N V_N} \frac{1}{|x|^{N-2}}, \quad N \geq 3.$$

**Theorem 8.4.3** *Let $N \geq 2$. In $\mathcal{D}^*(\mathbb{R}^N)$, we have*

$$-\Delta E_N = \delta.$$

**Proof** Define $v(x) = w(|x|)$. Since

$$\Delta v = w'' + (N-1)w'/|x|,$$

it is easy to verify that on $\mathbb{R}^N \setminus \{0\}$, $\Delta E_N = 0$. It is clear that $E_N \in L^1_{\text{loc}}(\mathbb{R}^N)$.

Let $u \in \mathcal{D}(\mathbb{R}^N)$ and $R > 0$ be such that spt $u \subset B(0, R)$. We have to verify that

$$-u(0) = \int_{\mathbb{R}^N} E_N \Delta u \, dx = \lim_{\varepsilon \to 0} \int_{\varepsilon < |x| < R} E_N \Delta u \, dx.$$

We obtain using the divergence theorem that

$$f(\varepsilon) = \int_{\varepsilon < |x| < R} (E_N \Delta u - u \Delta E_N) \, dx = \int_{\partial B(0,\varepsilon)} \left( u \nabla E_N \cdot \frac{\gamma}{|\gamma|} - E_N \nabla u \cdot \frac{\gamma}{|\gamma|} \right) d\gamma.$$

By a simple computation,

$$\int_{\partial B(0,\varepsilon)} \nabla E_N \cdot \frac{\gamma}{|\gamma|} = -1, \quad \lim_{\varepsilon \to 0} \int_{\partial B(0,\varepsilon)} E_N d\gamma = 0,$$

so that $\lim_{\varepsilon \to 0} f(\varepsilon) = -u(0)$.                                                                       □

**Definition 8.4.4** Let $f, g \in \mathcal{D}^*(\Omega)$. By definition, $f \leq g$ if for every $u \in \mathcal{D}(\Omega)$ such that $u \geq 0$, $\langle f, u \rangle \leq \langle g, u \rangle$.

**Theorem 8.4.5 (Kato's Inequality)** Let $g \in L^1_{\text{loc}}(\Omega)$ be such that $\Delta g \in L^1_{\text{loc}}(\Omega)$. Then

$$(\text{sgn } g) \Delta g \leq \Delta |g|.$$

**Proof** Let $u \in \mathcal{D}(\Omega)$ and $\omega \subset\subset \Omega$ be such that $u \geq 0$ and spt $u \subset \omega$. Define $g_n = \rho_n * g$, and for $\varepsilon > 0$, $f_\varepsilon(t) = (t^2 + \varepsilon^2)^{1/2}$. Since $g_n \to g$ in $L^1(\omega)$, we can assume, passing if necessary to a subsequence, that $g_n \to g$ almost everywhere on $\omega$.

For all $\varepsilon > 0$ and for $n$ large enough, we have

$$\int_\Omega f'_\varepsilon(g_n)(\Delta g_n)u \, dx \leq \int_\Omega (\Delta f_\varepsilon(g_n))u \, dx = \int_\Omega f_\varepsilon(g_n) \Delta u \, dx.$$

When $n \to \infty$, we find that

$$\int_\Omega f'_\varepsilon(g)(\Delta g)u \, dx \leq \int_\Omega f_\varepsilon(g) \Delta u \, dx.$$

When $\varepsilon \downarrow 0$, we obtain

$$\int_{\Omega} (\operatorname{sgn} g)(\Delta g)u \, dx \leq \int_{\Omega} |g| \, \Delta u \, dx. \qquad \square$$

## 8.5 Comments

The notion of polarization of sets appeared in 1952, in a paper by Wolontis [87]. Polarizations of functions were first used by Baernstein and Taylor to approximate symmetrization of functions on the sphere in the remarkable paper [3]. The explicit approximation of Schwarz's symmetrization by polarizations is due to Van Schaftingen [84]. See [73, 85] for other aspects of polarizations. The proof of Proposition 8.3.4 uses a device of Alberti [2]. The notion of symmetrization, and more generally, the use of reflections to prove symmetry, goes back to Jakob Steiner [79].

The elegant proof of Theorem 8.3.18 is due to O.S. Rothaus, *J. Funct. Anal. 64 (1985) 296–313.*

## 8.6 Exercises for Chap. 8

1. Let $u \in C(\Omega)$. The *spherical means* of $u$ are defined on $D$ by

$$S(x, r) = (N V_N)^{-1} \int_{\mathbb{S}^{N-1}} u(x + r\sigma)d\sigma.$$

Verify that when $u \in C^2(\Omega)$,

$$\lim_{r \downarrow 0} \frac{2N}{r^2}[S(x, r) - u(x)] = \Delta u(x).$$

2. Let $u \in C(\Omega)$ be such that for every $(x, r) \in D$, $u(x) = M(x, r)$. Then for every $x \in \Omega_n$, $\rho_n * u = u$.

   The argument is due to A. Ponce:

$$\rho_n * u(x) = \int_{\mathbb{R}^N} \rho_n(x - y)u(y)dy = \int_0^\infty dt \int_{\rho(x-y)>t} u(y)dy$$

$$= u(x) \int_0^\infty dt \int_{\rho(x-y)>t} dy = u(x).$$

3. (Weyl's theorem.) Let $u \in L^1_{\text{loc}}(\Omega)$. The following properties are equivalent:

   (a) $u$ is harmonic;

(b)  for almost all $x \in \Omega$ and for all $0 < r < d(x, \partial\Omega)$, $u(x) = M(x, r)$;

(c)  there exists $v \in C^\infty(\Omega)$, almost everywhere equal to $u$, such that $\Delta v = 0$.

4.  Let $u \in C^2(\Omega)$ be a harmonic function. Assume that $u \geq 0$ on $B[0, R] \subset \Omega$. Then for every $0 < r < R$ and $|y| < R - r$, we have

$$|u(y) - u(0)| \leq \frac{1}{r^N V_N} \int_{r-|y|<|x|<r+|y|} u(x)dx$$

$$= \frac{(r + |y|)^N - (r - |y|)^N}{r^N} u(0).$$

*Hint*: Use the mean-value property.

5.  (Liouville's theorem.) Let $u \in C^\infty(\mathbb{R}^N)$ be a harmonic function, bounded from below on $\mathbb{R}^N$. Then $u$ is constant.

6.  Let $\Omega$ be an open connected subset of $\mathbb{R}^N$, and let $u \in C^\infty(\Omega)$ be a harmonic function such that for some $x \in \Omega$, $u(x) = \inf_\Omega u$. Then $u$ is constant.

7.  If $u \in \mathcal{D}(]0, \pi[)$, then

$$\int_0^\pi \left|\frac{du}{dx}\right|^2 - u^2 dx = \int_0^\pi \left|\frac{du}{dx} - \frac{\cos x}{\sin x}u\right|^2 dx.$$

Hence

$$\min_{\substack{u \in H_0^1(]0,\pi[) \\ \|u\|_2 = 1}} \int_0^\pi \left|\frac{du}{dx}\right|^2 dx = 1.$$

8.  (Min–max principle.) For every $n \geq 1$,

$$\lambda_n = \min_{V \in \mathcal{V}_n} \max_{\substack{u \in V \\ \|u\|_2 = 1}} \int_\Omega |\nabla u|^2 dx,$$

where $\mathcal{V}_n$ denotes the family of all $n$-dimensional subspaces of $H_0^1(\Omega)$.

9.  Let us recall that

$$\lambda_1(G) = \inf\left\{\|\nabla u\|_2^2 / \|u\|_2^2 : u \in W_0^{1,2}(G) \setminus \{0\}\right\}.$$

Let $\Omega$ be an open subset of $\mathbb{R}^M$, and $\omega$ an open subset of $\mathbb{R}^N$. Then:

(a)  $\lambda_1(\Omega \times \omega) = \lambda_1(\Omega) + \lambda_1(\omega)$;

(b)  $\lambda_1(\mathbb{R}^N) = 0$;

(c)  $\lambda_1(\Omega \times \mathbb{R}^N) = \lambda_1(\Omega)$.

10. Define $u \in \mathcal{D}_+(\mathbb{R}^N)$ such that for every $y \in \mathbb{R}^N$, $\tau_y u \neq u^*$, and for $1 \leq p < \infty$, $||\nabla u||_p = ||\nabla u^*||_p$. *Hint*: Consider two functions $v$ and $w$ such that $v = v^*$, $w = w^*$, $v \equiv 1$ on $B(0, 1)$, and spt $w \subset B[0, 1/2]$, and define $u = v + \tau_y w$.

11. (Hardy–Littlewood inequality.) Let $1 < p < \infty$, $u \in L^p_+(\mathbb{R}^N)$, and $v \in L^{p'}_+(\mathbb{R}^N)$. Then

$$\int_{\mathbb{R}^N} u\, v\, dx \leq \int_{\mathbb{R}^N} u^* v^* dx.$$

12. Let $1 \leq p < \infty$ and $u, v \in L^p_+(\mathbb{R}^N)$. Then

$$||u + v||_p \leq ||u^* + v^*||_p.$$

*Hint*: Assume first that $p > 1$. Observe that

$$||u + v||_p = \sup_{\substack{w \in L^{p'} \\ ||w||_{p'}}} \int_{\mathbb{R}^N} (u + v)w\, dx.$$

13. Let $\Omega$ be a domain in $\mathbb{R}^N$ invariant under rotations. A function $u : \Omega \to \mathbb{R}$ is *foliated Schwarz's symmetric* with respect to $e \in \mathbb{S}^{N-1}$ if $u(x)$ depends only on $(r, \theta) = (|x|, \cos^{-1}(\frac{x}{|x|} \cdot e))$ and is decreasing in $\theta$.

Let $e \in \mathbb{S}^{N-1}$. We denote by $\mathcal{H}_e$ the family of closed half-spaces $H$ in $\mathbb{R}^N$ such that $0 \in \partial H$ and $e \in H$.

Prove that a function $u : \Omega \to \mathbb{R}$ is foliated Schwarz's symmetric with respect to $e$ if and only if for every $H \in \mathcal{H}_e$, $u^H = u$.

14. Let $u \in L^p(\mathbb{R}^N)(1 \leq p < \infty)$, and let the closed affine half-space $H \subset \mathbb{R}^N$ be such that $u^H = u$. Then, for every $n \geq 1$, $(\rho_n * u)^H = \rho_n * u$.

*Hint*. For every $x, y \in H$, we have

$$|x - y| = |\sigma_H(x) - \sigma_H(y)| \leq |x - \sigma_H(y)| = |\sigma_H(x) - y|.$$

Hence we obtain, for every $x \in H$,

$$\rho_n * u(x) - \rho_n * u(\sigma_H(x))$$
$$= \int_H \left[ u(y) - u(\sigma_H(y)) \right] \left[ \rho_n(x - y) - \rho_n(\sigma_H(x) - y) \right] dy \geq 0.$$

15. Let $u \in L^p(\mathbb{R}^N)(1 \leq p < \infty)$ be such that, for all $H \in \mathcal{H}$, $u^H = u$. Then $u \geq 0$ and $u = u^*$.

# Chapter 9
# Appendix: Topics in Calculus

## 9.1 Change of Variables

Our basic tool in this appendix is the following version of the *implicit function theorem*.

**Theorem 9.1.1** *Let $U$ be an open subset of $\mathbb{R}^N$, $\varphi \in C^1(U)$, and $a = (a', a_N) \in U$ such that $\partial_N \varphi(a) \neq 0$. Then there exist $r > 0$, $R > 0$ and*

$$\beta \in C^1\left(B(a', R) \times ]\varphi(a) - r, \varphi(a) + r[\right)$$

*such that, for $|x' - a'| < R$, $|t - \varphi(a)| < r$, we have*

$$\varphi(x', x_N) = t \iff x_N = \beta(x', t),$$

*and the set*

$$U_a = \left\{(x', \beta(x', t)) : |x' - a'| < R, |t - \varphi(a)| < r\right\}$$

*is an open neighborhood of $a$.*

**Definition 9.1.2** Let $U$ and $\omega$ be open subsets of $\mathbb{R}^N$. A diffeomorphism $f : U \to \omega$ is a continuously differentiable bijective mapping such that, for every $x \in U$,

$$J_f(x) = \det Df(x) \neq 0.$$

**Theorem 9.1.3** *Let $f : U \to \omega$ be a diffeomorphism and $u \in \mathcal{K}(\omega)$. Then $u(f) \in \mathcal{K}(U)$ and*

© Springer Nature Switzerland AG 2022
M. Willem, *Functional Analysis*, Cornerstones,
https://doi.org/10.1007/978-3-031-09149-0_9

$$\int_U u\big(f(x)\big)|J_f(x)|dx = \int_\omega u(y)dy. \quad (*)$$

**Lemma 9.1.4** *Formula* $(*)$ *is valid when* $N = 1$.

**Proof** We can assume that $U =\,]a, b[$. Then by the fundamental theorem of calculus, we have

$$\int_a^b u\big(f(x)\big)Df(x)dx = \int_{f(a)}^{f(b)} u(y)dy.$$

If $Df > 0$, then $\omega =\,]f(a), f(b)[$. If $Df < 0$, then $\omega =\,]f(b), f(a)[$. Hence formula $(*)$ is valid.                                             □

**Proof of Theorem 9.1.3** We will use induction on $N$. By Lemma 9.1.4, formula $(*)$ is valid when $N = 1$. Assume that $(*)$ is valid in dimension $N - 1$. Let $a \in U$. Since $f$ is a diffeomorphism, $\nabla f_N(a) \neq 0$. After a permutation of variables, we can assume that $\partial_N f_N(a) \neq 0$. Let $r > 0$, $R > 0$, $\beta$, and $U_a$ be given by Theorem 9.1.1 applied to $\varphi = f_N$. We factorize $f = (f', f_N)$ as $f = h(g)$ on $U_a$ by

$$g(x', x_N) = \big(x', f_N(x', x_N)\big), h(x', t) = \big(\Phi_t(x'), t\big),$$

where $\Phi_t(x') = f'(x', \beta(x', t))$.
    We assume that $u \in \mathcal{K}(f(U_a))$. Since $f = h(g)$ on $U_a$, we have that $Df = Dh(g)Dg$ and $J_f = J_h(g)J_g$. We define $v = u(h)|J_h|$, so that

$$\int_{U_a} u\big(f(x)\big)|J_f(x)|dx = \int u\big(h(g(x))\big)|J_h\big(g(x)\big)| \, |J_g(x)|dx$$

$$= \int v\big(g(x)\big)|J_g(x)|dx.$$

Fubini's theorem and Lemma 9.1.4 imply that

$$\int v\big(g(x)\big)|J_g(x)|dx = \int dx' \int v\big(x', f_N(x', x_N)\big)|\partial_N f_N(x', x_N)|dx_N$$

$$= \int dx' \int v(x', t)dt$$

$$= \int u\big(h(x', t)\big)|J_h(x', t)|dx'dt.$$

It follows from Fubini's theorem and the induction assumption that

$$\int u\big(h(x',t)\big)|J_h(x',t)|dx'dt = \int dt \int u(\Phi_t,x'),t)|J_{\Phi_t}(x')|dx'$$

$$= \int dt \int u(y',t)dy'$$

$$= \int_{f(U_a)} u(y)dy.$$

Hence formula (∗) is valid when $u \in \mathcal{K}(f(U_a))$.

Let $u \in \mathcal{K}(\omega)$. The Borel–Lebesgue theorem implies the existence of a finite covering of the compact set $f^{-1}(\text{spt } u)$ by open subsets $(U_{a_j})$ given by the implicit function theorem as before. There exists also a continuous partition of unity $(\psi_j)$ subordinate to the covering of spt $u$ by $(f(U_{a_j}))$. Since $u = \sum_j \psi_j u$, it is easy to conclude the proof. □

## 9.2 Surface Integrals

In this section, we assume that $U$ is an open subset of $\mathbb{R}^N$, $u \in C^1(U)$, and $f \in \mathcal{K}(U)$. Our goal is to prove that, under some assumptions,

$$\int_U f(x)|\nabla u(x)|dx = \int_{\mathbb{R}} dt \int_{u=t} f(y)dy, \qquad (*)$$

$$\frac{d}{dt}\int_{u<t} f(x)|\nabla u(x)|dx = \int_{u=t} f(y)dy, \qquad (**)$$

where

$$(u < t) = \{u < t\} = \{x \in U : u(x) < t\}$$

and

$$(u = t) = \{u = t\} = \{y \in U : u(y) = t\}.$$

**Definition 9.2.1** Let $a \in U$ be such that $\nabla u(a) \neq 0$. After a permutation of variables, we can assume that $\partial_N u(a) \neq 0$. Let $r > 0$, $R > 0$, $\beta$, and $U_a$ be given by Theorem 9.1.1 applied to $\varphi = u$. Let $f \in \mathcal{K}(U_a)$. We define for $|t - u(a)| < r$,

$$\int_{u=t} f(y)dy = \int_{B(a',R)} f(x',\beta(x',t))\sqrt{1+|\nabla_{x'}\beta(x',t)|^2}dx'.$$

**Lemma 9.2.2** *Let* $f \in \mathcal{K}(U_a)$. *Then we have that*

$$\int_{U_a} f(x)|\nabla u(x)|dx = \int_{u(a)-r}^{u(a)+r} dt \int_{u=t} f(\gamma)d\gamma.$$

*Moreover formula* (**) *is valid for* $|t - u(a)| < r$.

**Proof** We define the change of variables $h(x', t) = (x', \beta(x', t))$, and we choose

$$u(a) - r \le b < c \le u(a) + r.$$

We obtain, using Theorem 9.1.3 and Fubini's theorem, that

$$\int_{b<u<c} f(x)|\nabla u(x)|dx = \int_b^c dt \int_{B(a',R)} f\big(x', \beta(x', t)\big)\big|\nabla u(x', \beta(x', t))\big|\,\big|\partial_t \beta(x', t)\big|dx'.$$

Since, by definition, $u(x', \beta(x', t)) = t$, it follows that

$$\nabla_{x'} u\big(x', \beta(x', t)\big) + \partial_N u\big(x', \beta(x' + t)\big)\nabla_{x'}\beta(x', t) = 0,$$
$$\partial_N u\big(x', \beta(x', t)\big)\partial_t \beta(x', t) = 1$$

and

$$\big|\nabla u(x', \beta(x', t))\big|^2 \big|\partial_t \beta(x', t)\big|^2 = 1 + \big|\nabla_{x'}\beta(x', t)\big|^2.$$

Hence we obtain

$$\int_{b<u<c} f(x)|\nabla u(x)|dx = \int_b^c dt \int_{B(a',R)} f\big(x', \beta(x', t)\big)\sqrt{1 + |\nabla_{x'}\beta(x', t)|^2}dx'$$
$$= \int_b^c dt \int_{u=t} f(\gamma)d\gamma.$$

In particular, for $|t - u(a)| < r$, the fundamental theorem of calculus implies that

$$\lim_{\epsilon \downarrow 0} \frac{1}{\epsilon} \int_{t<u<t+\epsilon} f(x)|\nabla u(x)|dx = \int_{u=t} f(\gamma)d\gamma = \lim_{\epsilon \downarrow 0} \frac{1}{\epsilon} \int_{t-\epsilon<u<t} f(x)|\nabla u(x)|dx. \quad \square$$

**Definition 9.2.3** A regular value of $u \in C^1(U)$ is a real number $c$ such that

$$x \in U \text{ and } u(x) = c \Rightarrow \nabla u(x) \neq 0.$$

**Definition 9.2.4** Let $f \in \mathcal{K}(U)$ and let $c$ be a regular value of $u \in C^1(U)$. The Borel–Lebesgue theorem implies the existence of a finite covering of the compact set spt $f \cap \{u = c\}$ by open subsets $(U_{a_j})_{1 \leq j \leq k}$ given by Definition 9.2.1. There exists a continuous partition of unity $(\psi_j)$ subordinate to the covering $(U_{a_j})_{1 \leq j \leq k}$. By definition

$$\int_{u=c} f(\gamma)d\gamma = \sum_{j=1}^{k} \int_{u=c} \psi_j(\gamma)f(\gamma)d\gamma.$$

Let us prove that the surface integral $\int_{u=c} f(\gamma)d\gamma$ depends only on $f$, $u$, and $c$.

**Theorem 9.2.5** *Let $f \in \mathcal{K}(U)$, and let $c$ be a regular value of $u \in C^1(U)$. Then formula* (∗∗) *is valid at $t = c$.*

**Proof** Let us define $\psi_0 = 1 - \sum_{j=1}^{k} \psi_j$. Since $\psi_0 = 0$ on a neighborhood of spt $f \cap \{u = c\}$, it follows from Lemma 9.2.2 and Definition 9.2.4 that

$$\frac{d}{dt}\Big|_{t=c} \int_{u<t} f(x)|\nabla u(x)|dx = \sum_{j=0}^{k} \frac{d}{dt}\Big|_{t=c} \int_{u<t} \psi_j(x)f(x)|\nabla u(x)|dx$$

$$= \sum_{j=1}^{k} \int_{u=c} \psi_j(\gamma)f(\gamma)d\gamma$$

$$= \int_{u=c} f(\gamma)d\gamma. \qquad \square$$

**Proposition 9.2.6** *Let $f \in \mathcal{K}(U)$, and let $u \in C^1(U)$ be such that, for every $x \in U$, $\nabla u(x) \neq 0$. Then formula* (∗) *is valid.*

**Proof** The Borel–Lebesgue theorem implies the existence of a finite covering of the compact set spt $f$ by open subsets $(U_{a_j})$ given by Definition 9.2.1. There exists a continuous partition of unity $\psi_j$ subordinate to the finite covering $(U_{a_j})$. It follows from Lemma 9.2.2 and Definition 9.2.4 that

$$\int_U f(x)|\nabla u(x)|dx = \sum_j \int_{U_{a_j}} \psi_j(x)f(x)|\nabla u(x)|dx$$

$$= \sum_j \int_{\mathbb{R}} dt \int_{u=t} \psi_j(\gamma)f(\gamma)d\gamma$$

$$= \int_{\mathbb{R}} dt \sum_j \int_{u=t} \psi_j(\gamma)f(\gamma)d\gamma = \int_{\mathbb{R}} dt \int_{u=t} f(\gamma)d\gamma. \qquad \square$$

## 9.3 The Morse–Sard Theorem

The *Morse–Sard theorem* ensures that almost all values of a smooth function are regular.

**Theorem 9.3.1** *Let $U$ be an open subset of $\mathbb{R}^N$, let $u \in C^\infty(U)$, and define*

$$C_1 = \{x \in U : \nabla u(x) = 0\}.$$

*Then Lebesgue's measure of $u(C_1)$ is equal to 0.*

**Lemma 9.3.2** *Let $u \in C^{N+1}(U)$ and define*

$$C_N = \{x \in U : for\ |\alpha| \leq N,\ \partial^\alpha u(x) = 0\}.$$

*Then Lebesgue's measure of $u(C_N)$ is equal to 0.*

**Proof** Since $U$ is covered by a countable family of closed cubes, it suffices to prove that $u(C_N \cap K)$ is negligible, where $K = B_\infty[x, r/2] \subset U$.

By definition of $C_N$, Taylor's formula implies the existence of $c \geq 0$ such that for every $x \in C_N \cap K$ and every $y \in K$,

$$\left|u(x) - u(y)\right| \leq c \|x - y\|_\infty^{N+1}.$$

We divide $K$ into $2^{jN}$ cubes with edge $r/2^j$. Then $u(C_N \cap K)$ is contained in at most $2^{jN}$ intervals of length $2c(r/2^j)^{N+1}$. We conclude that

$$m\big((C_N \cap K)\big) \leq 2^{jN} 2c(r/2^j)^{N+1} = 2cr^{N+1}/2^j \to 0, j \to \infty. \qquad \square$$

**Proof of Theorem 9.3.1** We will use induction on $N$. By Lemma 9.3.2, the theorem is valid when $N = 1$.

Assume that the theorem is valid in dimension $N - 1$, and define

$$C_k = \big\{x \in U : \text{for every } |\alpha| \leq k,\ \partial^\alpha u(x) = 0\big\}.$$

By Lemma 9.3.2 it suffices to prove that $u(C_k \backslash C_{k+1})$ is negligible for $1 \leq k \leq N - 1$.

Let $a \in \big(C_k \backslash C_{k+1}\big)$. By definition, there exist $\alpha \in \mathbb{N}^N$ and $1 \leq j \leq N$ such that $|\alpha| = k$, $\partial^\alpha u(a) = 0$, and $\partial_j \partial^\alpha u(a) \neq 0$. After a permutation of variables, we can assume that $\partial_N \partial^\alpha u(a) \neq 0$. Let $r > 0$, $R > 0$, $\beta$, and $U_a$ be given by Theorem 9.1.1 applied to $\varphi = \partial^\alpha u$. Since $u \in C^\infty(U)$, it follows that $\beta \in C^\infty\big(B(a', R) \times ]-r, r[\big)$.

Let us define $v$ on $B(a', R)$ by $v(x') = u\big(x', \beta(x', 0)\big)$. It follows from the induction assumption that

$$m\{v(x') : x' \in B(a', R) \text{ and } \nabla v(x') = 0\} = 0 \quad (*)$$

Let $x \in C_k \cap U_a$. Since, by definition, $\varphi(x) = \partial^\alpha u(x) = 0$ and $\nabla u(x) = 0$, we obtain $\beta(x', 0) = x_N$ and

$$v(x') = u(x', \beta(x', 0)) = u(x)$$
$$\nabla v(x') = \nabla_{x'} u(x', \beta(x', 0)) + \partial_N u(x', \beta(x', 0)) \nabla_{x'} \beta(x', 0)$$
$$= \nabla_{x'} u(x) + \partial_N u(x) \nabla_{x'} \beta(x', 0) = 0.$$

We deduce from $(*)$ that

$$m\big(u(C_k \cap U_a)\big) = 0.$$

Let us define, for $n \geq 1$,

$$K_n = \{x \in C_k : d(x, C_{k+1}) \geq 1/n, d(x, \partial U) \geq 1/n, |x| \leq 1/n\}.$$

The Borel–Lebesgue theorem implies the existence of a finite covering of the compact set $K_n$ by open subsets $(U_{a_j})$ satisfying $m(u(C_k \cap U_{a_j})) = 0$. It follows that, for $n \geq 1$, $m(u(K_n)) = 0$. We conclude that

$$m\big(u(C_k \backslash C_{k+1})\big) = m\Big(u\Big(\bigcup_{n=1}^{\infty} K_n\Big)\Big) = 0. \qquad \square$$

The following theorem is a version of *the coarea formula*.

**Theorem 9.3.3** *Let $U$ be an open subset of $\mathbb{R}^N$, let $u \in C^\infty(U)$, and let $f \in C(U)$ be such that $\int_U |f| |\nabla u| dx < \infty$. Then*

$$\int_U f(x) |\nabla u(x)| dx = \int_{\mathbb{R}} dt \int_{u=t} f(\gamma) d\gamma.$$

***Proof*** We define

$$C = \{x \in U : \nabla u(x) = 0\}$$

and

$$\omega_n = \{x \in U : d(x, C) > 1/n, d(x, \partial U) > 1/n \text{ and } |x| < n\}.$$

For every $n \geq 1$, there exists $\varphi_n \in \mathcal{D}(\omega_{n+1})$ such that $0 \leq \varphi_n \leq 1$ and $\varphi_n = 1$ on $\omega_n$.

Proposition 9.2.6 implies that

$$\int_U f\varphi_n|\nabla u|dx = \int_{\omega_{n+1}} f\varphi_n|\nabla u|dx = \int_{\mathbb{R}} dt \int_{u|_{\omega_{n+1}}=t} f\varphi_n d\gamma.$$

If $t$ is a regular value of $u$, it is clear that

$$\int_{u|_{\omega_{n+1}}=t} f\varphi_n d\gamma = \int_{u=t} f\varphi_n d\gamma.$$

Hence the Morse–Sard theorem implies that

$$\int_U f\varphi_n|\nabla u|dx = \int_{\mathbb{R}} dt \int_{u=t} f\varphi_n d\gamma,$$

where the surface integral is defined only when $t$ is a regular value of $u$. It follows from the definition of $\varphi_n$ that $\varphi_n \uparrow \chi_{U\setminus C}$. We can assume that $f \geq 0$. We conclude, using Levi's theorem, that

$$\int_U f|\nabla u|dx = \lim_{n\to\infty} \int_U f\varphi_n|\nabla u|dx = \lim_{n\to\infty} \int_{\mathbb{R}} dt \int_{u=t} f\varphi_n d\gamma = \int_{\mathbb{R}} dt \int_{u=t} f d\gamma.$$

$\square$

## 9.4   The Divergence Theorem

An open subset of $\mathbb{R}^N$ is *smooth* if its boundary is a smooth manifold.

**Definition 9.4.1** Let $m \geq 1$. The open subset $\Omega$ of $\mathbb{R}^N$ is of class $C^m$ if there exists $\varphi \in C^m(\mathbb{R}^N)$ such that

(a) $\Omega = \left\{x \in \mathbb{R}^N : \varphi(x) < 0\right\}$;
(b) $\Gamma = \partial\Omega = \left\{\gamma \in \mathbb{R}^N : \varphi(\gamma) = 0\right\}$;
(c) for every $\gamma \in \Gamma$, $\nabla\varphi(\gamma) \neq 0$.

The exterior normal at $\gamma \in \Gamma$ is defined by

$$n(\gamma) = \nabla\varphi(\gamma)/\left|\nabla\varphi(\gamma)\right|.$$

The boundary integral is the elementary integral defined on $\mathcal{K}(\mathbb{R}^N)$ by

$$\int_\Gamma u\, d\gamma = \int_{\varphi=0} u\, d\gamma.$$

*Notation* Let us define $\eta \colon \mathbb{R} \to \mathbb{R}$ by

$$\eta(t) = 1, \qquad t \leq -1,$$
$$= -t, \qquad -1 \leq t \leq 0,$$
$$= 0, \qquad t \geq 0.$$

**Lemma 9.4.2** *For every $n \geq 1$, we have that $(\rho_n * \eta)' = -\rho_n * \chi_{]-1,0[}$, $0 \leq \rho_n * \eta \leq 1$, and $0 \leq \rho_n * \chi_{]-1,0[} \leq 1$.*

**Proof** Proposition 4.3.6 implies that, for every $x \in \mathbb{R}$,

$$(\rho_n * \eta)'(x) = \int_{\mathbb{R}} \rho_n'(x - y)\eta(y)dy$$
$$= \int_{-\infty}^{-1} \rho_n'(x - y)dy - \int_{-1}^{0} \rho_n'(x - y)y\,dy$$
$$= -\int_{-1}^{0} \rho_n(x - y)dy.$$

Since $0 \leq \eta \leq 1$, we obtain $0 \leq \rho_n * \eta \leq \rho_n * 1 = 1$. The case of $\chi_{]-1,0[}$ is identical. $\qquad\square$

**Theorem 9.4.3 (Divergence Theorem)** *Let $\Omega$ be an open subset of $\mathbb{R}^N$ of class $C^1$, and let $v \in C^1(\mathbb{R}^N; \mathbb{R}^N) \cap \mathcal{K}(\mathbb{R}^N; \mathbb{R}^N)$. Then*

$$\int_{\Omega} \operatorname{div} v\,dx = \int_{\Gamma} v \cdot n\,d\gamma.$$

**Proof** Lemmas 6.1.1 and 9.4.2 imply that, for every $\epsilon > 0$ and $n \geq 1$,

$$\int_{\mathbb{R}^N} \rho_n * \eta(\varphi/\epsilon) \operatorname{div} v\,dx = \frac{1}{\epsilon} \int_{\mathbb{R}^N} \rho_n * \chi_{]-1,0[}(\varphi/\epsilon)\nabla\varphi \cdot v\,dx.$$

Using the regularization theorem and Lebesgue's dominated convergence theorem, we obtain, for every $\epsilon > 0$,

$$\int_{\mathbb{R}^N} \eta(\varphi/\epsilon) \operatorname{div} v\,dx = \frac{1}{\epsilon} \int_{\mathbb{R}^N} \chi_{]-1,0[}(\varphi/\epsilon)\nabla\varphi \cdot v\,dx.$$

Using again Lebesgue's dominated convergence theorem and Theorem 9.2.5, we conclude that

$$\int_{\varphi<0} \operatorname{div} v\,dx = \lim_{\substack{\epsilon \to 0 \\ \epsilon > 0}} \int_{\mathbb{R}^N} \eta(\varphi/\epsilon) \operatorname{div} v\,dx = \lim_{\substack{\epsilon \to 0 \\ \epsilon > 0}} \frac{1}{\epsilon} \int_{-\epsilon<\varphi<0} v \cdot \nabla\varphi\,dx = \int_{\Gamma} v \cdot \frac{\nabla\varphi}{|\nabla\varphi|}d\gamma. \quad\square$$

## 9.5   Comments

The proofs of Theorem 9.1.3, Lemma 9.2.2, and Theorem 9.3.1 depend only on the implicit function theorem for one equation and one dependent variable (Theorem 9.1.1). A direct proof of this result is given in the book of Krantz and Parks on the implicit function theorem ([41], Theorem 3.2.1).

The change of variable formula for a double integral was discovered by L. Euler (*De formulis integralibus duplicatis, Novi Comm. acad. Scient. Petropolitanae*, 14 (1769) 72–103). The proof consists in factorizing the change of variables leaving one variable fixed and transforming the other. A more recent version is given in *Differential and Integral Calculus* by R. Courant, vol. II, p. 247. The same book contains the coarea formula (for regular values) under the name of *resolution of multiples integrals* (p. 302).

The proof of the Morse–Sand theorem for smooth functions (Theorem 9.3.1) is due to Milnor. The short proof of the divergence theorem in Sect. 9.4 was inspired by Example 7.2, Chapter 3, in the book [40] by Krantz and Parks.

# Chapter 10
# Epilogue: Historical Notes on Functional Analysis

*Differentiae et summae sibi reciprocae sunt, hoc est summa
differentiarum seriei est seriei terminus, et differentia
summarum seriei est ipse seriei terminus, quorum illud ita
enuntio: ∫ dx aequ. x; hoc ita: d ∫ x aequ. x.*

G. Leibniz

## 10.1   Integral Calculus

In a concise description of mathematical methods, Henri Lebesgue underlined the importance of definitions and axioms (see [47]):

When a mathematician foresees, more or less clearly, a proposition, instead of having recourse to experiment like the physicist, he seeks a logical proof. For him, logical verification replaces experimental verification. In short, he does not seek to discover new materials but tries to become aware of the richness that he already unconsciously possesses, which is built in the definitions and axioms. Herein lies the supreme importance of these definitions and axioms, which are indeed subjected logically only to the condition that they be compatible, but which could lead only to a purely formal science, void of meaning, if they had no relationship to reality.

Leibniz conceived integration as the reciprocal of differentiation:

$$\int dx = d \int x = x.$$

The computation of the integral of $f$ is reduced to the search for its *primitive*, solution of the differential equation

$$F' = f.$$

© Springer Nature Switzerland AG 2022
M. Willem, *Functional Analysis*, Cornerstones,
https://doi.org/10.1007/978-3-031-09149-0_10

The textbooks by Cauchy, in particular the *Analyse algébrique* (1821) (see [7]) and the *Résumé des leçons données à l'Ecole Royale Polytechnique sur le calcul infinitésimal* (1823), opened a new area in analysis. Cauchy was the first to consider the problem of existence of primitives:

> In integral calculus, it seemed necessary to me to demonstrate in general the existence of *integrals* or *primitive functions* before giving their various properties. In order to reach this, it was necessary to establish the notion of *integral between two given limits* or *definite integral*.

Cauchy defines and proves the existence of the integral of continuous functions:

> According to the preceding lecture, if one divides $X - x_0$ into infinitesimal elements $x_1 - x_0, x_2 - x_1 \cdots X - x_{n-1}$, the sum
>
> $$S = (x_1 - x_0) f(x_0) + (x_1 - x_2) f(x_1) + \cdots + (X - x_{n-1}) f(x_{n-1})$$
>
> will converge to a limit given by the definite integral
>
> $$\int_{x_0}^{X} f(x) dx.$$

So Cauchy proved the existence of primitives of continuous functions using integral calculus.

Though every continuous function has a primitive, Weierstrass proved in 1872 the existence of continuous nowhere differentiable functions. In a short note [44], Lebesgue proved the existence of primitives of continuous functions without using integral calculus. His proof is clearly functional-analytic.

In 1881 [37], Camille Jordan defined the functional space of *functions of bounded variation*, which he called functions of *limited oscillation*. His goal was to linearize Dirichlet's condition for the convergence of Fourier series:

> Let $x_1, \ldots, x_n$ be a series of values of $x$ between 0 and $\varepsilon$, and $y_1, \ldots, y_n$ the corresponding values of $f(x)$. The points $x_1, y_1; \ldots; x_n, y_n$ will form a polygon.
> Consider the differences
>
> $$y_2 - y_1, y_3 - y_2, \ldots, y_n - y_{n-1}.$$
>
> We will call the sum of the positive terms of this sequence the *positive oscillation* of the polygon; *negative oscillation* is the sum of the negative terms; *total oscillation* is the sum of those two partial oscillations in absolute value.
> Let us vary the polygon; two cases may occur:
>
> 1° The polygon may be chosen so that its oscillations exceed every limit.
> 2° For every chosen polygon, its positive and negative oscillations will be less than some fixed limits $P_\varepsilon$ and $N_\varepsilon$. We will say in that case that $F(x)$ is a function of *limited oscillation* in the interval from 0 to $\varepsilon$; $P_\varepsilon$ will be its *positive oscillation*; $N_\varepsilon$ its *negative oscillation*; $P_\varepsilon + N_\varepsilon$ its *total oscillation*.

This case will necessarily occur if $F(x)$ is the difference of two finite functions $f(x) - \varphi(x)$, because it is clear that the positive oscillation of the polygon will be $\underset{<}{=} f(\varepsilon) - f(0)$, and its negative oscillation $\underset{<}{=} \varphi(\varepsilon) - \varphi(0)$.

The converse is easy to prove. Indeed, it is easy to verify that

1° The oscillation of a function from 0 to $\varepsilon$ is equal to the sum of its oscillations from 0 to $x$ and from $x$ to $\varepsilon$, $x$ being any quantity between 0 and $\varepsilon$.

2° We have that $F(x) = F(0) + P_x - N_x$, $P_x$ and $N_x$ denoting the positive and the negative oscillations from 0 to $x$. But $F(0) + P_x$ and $N_x$ are finite functions nondecreasing from 0 to $\varepsilon$.

Hence Dirichlet's proof is applicable, without modification, to every function of bounded oscillation from $x = 0$ to $x = \varepsilon$, $\varepsilon$ being any finite quantity.

The functions of limited oscillations constitute a well-defined class, whose study could be of some interest.

Functions of bounded variation will play a fundamental role in the following domains:

(a) Convergence of Fourier series;
(b) Rectification of curves;
(c) Integration;
(d) Duality.

Let $u : [0, 1] \to \mathbb{R}$ be a continuous function. The length of the graph of $u$ is defined by

$$L(u) = \sup \left\{ \sum_{j=0}^{k} \left[ (a_{j+1} - a_j)^2 + \left( u(a_{j+1}) - u(a_j) \right)^2 \right]^{1/2} : \right.$$

$$\left. k \in \mathbb{N}, 0 = a_0 < a_1 < \ldots < a_{k+1} = 1 \right\}.$$

In 1887, in Volume III of the first edition of his *Cours d'Analyse* at the École Polytechnique, Jordan proved that $L(u)$ is finite if and only if $u$ is of bounded variation. The case of surfaces is much more delicate (see Sect. 10.3).

In 1894 [80], Stieltjes defined a deep generalization of the integral associated with an increasing function $\varphi$:

More generally, let us consider the sum

$$f(\xi_1)\big[\varphi(x_1) - \varphi(x_0)\big] + f(\xi_2)\big[\varphi(x_2) - \varphi(x_1)\big] + \ldots + f(\xi_n)\big[\varphi(x_n) - \varphi(x_{n-1})\big]. \quad \text{(A)}$$

It will still have a limit, which we shall denote by

$$\int_a^b f(u)d\varphi(u).$$

We will have only to consider some very simple cases like $f(u) = u^k$, $f(u) = \frac{1}{z+u}$, and there is no interest in giving to the function $f(u)$ its full generality. Thus it will suffice, as an example, to suppose the function $f(u)$ continuous, and then the proof presents no difficulty, and we have no need to develop it, since it is done as in the ordinary case of a definite integral.

It is easy to extend Stieltjes's definition to every function $\varphi$ of bounded variation. Stieltjes breaks the reciprocity between integral and derivative.

In 1903 [32], J. Hadamard characterized the *continuous linear functionals* on $C([a, b])$:

It is easy to reach this, following Weierstrass and Kirchhoff, and introducing a function $F(x)$, with a finite number of maxima and minima and such that

$$\int_{-\infty}^{+\infty} F(x)dx = 1;$$

e.g., $F(x) = \frac{1}{\sqrt{\pi}} e^{-x^2}$.

Starting then from the well-known identity

$$\lim_{\mu = \pm \infty} \mu \int_b^a f(x) F[\mu(x - x_0)]dx = f(x_0), \quad a < x_0 < b,$$

and assuming (as the authors quoted before) the operation $U$ to be continuous (in the sense of Bourlet), it will suffice to define

$$U[\mu F \mu(x - x_0)] = \Phi(x_0, \mu)$$

to show that our operation could be represented as

$$U[f(x)] = \lim_{\mu = \pm \infty} \int_a^b f(x) \Phi(x, \mu)dx.$$

In 1909 [61], F. Riesz discovered a representation depending on only one function:

In the present note, we shall develop a new analytic expression of the linear operation, containing only one generating function.

Given the linear operation $A[f(x)]$, we can determine a function of bounded variation $\alpha(x)$ such that for every continuous function $f(x)$, we have

$$A[f(x)] = \int_0^1 f(x)d\alpha(x).$$

Riesz's theorem asserts that every continuous linear functional on $C([0, 1])$ is representable by Stieltjes's integral.

## 10.2 Measure and Integral

> *Les notions introduites sont exigées par la solution d'un*
> *problème, et, en vertu de la seule présence parmi les notions*
> *antérieures, elles posent à leur tour de nouveaux problèmes.*
>
> Jean Cavaillès

In 1898, Emile Borel defined the measure of sets in his *Leçons sur la théorie des fonctions*:

> The procedure that we have employed actually amounts to this: we have recognized that a definition of measure could be useful only if it had certain fundamental properties: we have stated these properties a priori, and we have used them to define the class of sets that we consider measurable.
>
> Those essential properties that we summarize here, since we shall use them, are the following: The measure of a sum of a denumerable infinity of sets is equal to the sum of their measures; the measure of the difference of two sets is equal to the difference of their measures; *the measure is never negative; every set with a nonzero measure is not denumerable.* It is mainly this last property that we shall use. Besides, it is explicitly understood that we speak of measures only for those sets that we called *measurable*.
>
> Of course, when we speak of the sum of several sets, we assume that every pair them have no common points, and when we speak of their difference, we assume that one set contains all the points of the other.

Following Lebesgue:

> The descriptive definition of measure stated by M. Borel is without doubt the first clear example of the use of actual infinity in mathematics.

However, Borel does not prove the existence of the measure!

Lebesgue's integral first appeared on 29 April 1901. In the note [42], Lebesgue proved the existence of Borel's measure as a restriction of Lebesgue's measure.

In the introduction of his thesis [43], Lebesgue stated his program:

> In this work, I try to give definitions as general and precise as possible of some of the numbers considered in Analysis: definite integral, length of a curve, area of a surface.

He formulated the *problem of the measure of sets*:

> We intend to assign to every bounded set a positive or zero number called its measure and satisfying the following conditions:
>
> 1. There exist sets with nonzero measure.
> 2. Two equal sets have equal measures.
> 3. The measure of the sum of a finite number or of a countable infinity of sets, without common points, is the sum of the measures of those sets.
>
> We will solve this *problem of measure* only for the sets that we will call measurable.

In his *Leçons sur l'intégration et la recherche des fonctions primitives* of 1904, see [45], Lebesgue formulated the *problem of integration*:

We intend to assign to every bounded function $f(x)$ defined on a finite interval $(a, b)$, positive, negative, or zero, a finite number $\int_a^b f(x)dx$, which we call the integral of $f(x)$ in $(a, b)$ and which satisfies the following conditions:

1. For every $a, b, h$, we have

$$\int_a^b f(x)dx = \int_{a+h}^{b+h} f(x - h)dx.$$

2. For every $a, b, c$, we have

$$\int_a^b f(x)dx + \int_b^c f(x)dx + \int_c^a f(x)dx = 0.$$

3.

$$\int_a^b [f(x) + \varphi(x)]dx = \int_a^b f(x)dx + \int_a^b \varphi(x)dx.$$

4. If we have $f \geqq 0$ and $b > a$, we also have

$$\int_a^b f(x)dx \geqq 0.$$

5. We have

$$\int_0^1 1 \times dx = 1.$$

6. If $f_n(x)$ increases and converges to $f(x)$, then the integral of $f_n(x)$ converges to the integral of $f(x)$.

Formulating the six conditions of the integration problem, we define the integral. This definition belongs to the class of those that could be called *descriptive*; in those definitions, we state the characteristic properties of the object we want to define. In the *constructive* definitions, we state which operations are to be done in order to obtain the object we want to define. Constructive definitions are more often used in Analysis; however, we use sometimes descriptive definitions; the definition of the integral, following Riemann, is constructive; the definition of primitive functions is descriptive.

In 1906, in his thesis [23], Maurice Fréchet tried to extend the fundamental notions of analysis to abstract sets:

In this Mémoire we will use an absolutely general point of view that encompass these different cases.

To this end, we shall say that a *functional operation* $U$ is defined on a set $E$ of elements of *every kind* (numbers, curves, points, etc.) when to every element $A$ of $E$ there corresponds a determined numerical value of $U : U(A)$. The search for properties of those operations constitutes the object of the *Functional Calculus*.

Fréchet defined *distance* which he called, in French, *écart*:

We can associate to every pair of elements $A, B$ a number $(A, B) \geq 0$, which we will call *the distance of the two elements* and which satisfies the following properties: (a) The distance $(A, B)$ is zero only if $A$ and $B$ are identical. (b) If $A, B, C$ are three arbitrary elements, we always have $(A, B) \le (A, C) + (C, B)$.

In [24], Fréchet defined *additive families of sets* and *additive functions of sets*:

*An additive family of sets* is a collection of sets such that:

1. If $E_1$, $E_2$ are two sets of this family, the set $E_1 - E_2$ of elements of $E_1$, if they exist and that are not in $E_2$, belongs also to the family.
2. If $E_1, E_2, \ldots$ is a denumerable sequence of sets of this family, their sum, i.e., the set $E_1 + E_2 + \cdots$ of elements belonging at least to one set of the sequence, belongs also to the family.

A set function $f(E)$ defined on an additive family of sets $\mathcal{F}$ is *additive* on $\mathcal{F}$ if $E_1, E_2, \ldots$ being a denumerable sequence of sets of $\mathcal{F}$ and *disjoint*, i.e., without pairwise common elements, we have

$$f(E_1 + E_2 + \ldots) = f(E_1) + f(E_2) + \cdots .$$

When the sequence is infinite, the second member has obviously to converge regardless of the order of the terms. Hence the series in the second member has to converge absolutely.

Fréchet defined the integral without using topology. *Additive functions of sets* will be called *measures*.

In [12], Daniell chose a different method. He introduced a space $\mathcal{L}$ of *elementary functions* and an *elementary integral*

$$\mathcal{L} \to \mathbb{R} : u \mapsto \int u \, d\mu$$

satisfying the axioms of linearity, positivity, and monotone convergence.

The two axiomatics are equivalent if to Daniell's axioms we add *Stone's axiom* (1948):

$$\text{for every } u \in \mathcal{L}, \min(u, 1) \in \mathcal{L},$$

or the axiom

$$\text{for every } u, v \in \mathcal{L}, uv \in \mathcal{L}.$$

The choice of primitive notions and axioms is rather arbitrary. There are no absolutely undefinable notions or unprovable propositions.

The axiomatization of integration by Fréchet opened the way to the axiomatization of probability by Kolmogorov in 1933. The unification of measure, integral, and probability was one the greatest scientific achievements of the twentieth century.

In his thesis [5], Banach defined the *complete normed spaces*:

There exists an operation, called norm (we shall denote it by the symbol $||X||$), defined in the field $E$, having as an image the set of real numbers and satisfying the following conditions:

$||X|| \ge 0,$

$||X|| = 0$ if and only if $X = \theta$,
$||a \cdot X|| = |a| \cdot ||X||$,
$||X + Y|| \leq ||X|| + ||Y||$.

If 1. $\{X_n\}$ is a sequence of elements of $E$, 2. $\lim\limits_{\substack{r \to \infty \\ p \to \infty}} ||X_r - X_p|| = 0$, there exists an element $X$ such that

$$\lim_{n \to \infty} ||X - X_n|| = 0.$$

Banach emphasized the efficiency of the axiomatic method:

> The present work intends to prove theorems valid for different functional fields, which I will specify in the sequel. However, in order not to be forced to prove them individually for every particular field, a tedious task, I chose a different way: I consider in some general way sets of elements with some axiomatic properties, I deduce theorems, and I prove afterward that the axioms are valid for every specific functional field.

The fundamental book of Banach [6], *Théorie des opérations linéaires*, was published in 1932. Banach deduces Riesz's representation theorem from the Hahn–Banach theorem.

The original proof of the Hahn–Banach theorem holds in every real vector space. Let $F : X \to \mathbb{R}$ be a positively homogeneous convex function, and let $f : Z \to \mathbb{R}$ be a linear function such that $f \leq F$ on the subspace $Z$ of $X$. By the well-ordering theorem, the set $X \setminus Z$ can be so ordered that each nonempty subset has a least element. It follows then, from Lemma 4.1.3, by transfinite induction, that there exists $g : X \to \mathbb{R}$ such that $g \leq F$ on $X$ and $g\big|_Z = f$.

Let us recall the *principle of transfinite induction* (see [72]). Let $\mathcal{B}$ be a subset of a well-ordered set $\mathcal{A}$ such that

$$\{y \in \mathcal{A} : y < x\} \subset \mathcal{B} \Rightarrow x \in \mathcal{B}.$$

Then $\mathcal{B} = \mathcal{A}$.

In set theory, the well-ordering theorem is equivalent to the axiom of choice and to Zorn's lemma. In 1905, Vitali proved the existence of a subset of the real line that is not Lebesgue measurable. His proof depends on the axiom of choice.

## 10.3   Differential Calculus

*L'activité des mathématiciens est une activité expérimentale.*

Jean Cavaillès

Whereas the integral calculus transforms itself into an *axiomatic theory*, the differential calculus fits into the *general theory* of distributions.

The fundamental notions are

- Weak solutions;
- Weak derivatives;
- Functions of bounded variation;
- Distributions.

In [60], Poincaré defined the notion of *weak solution* of a boundary value problem:

Let $u$ be a function satisfying the following conditions:

$$\frac{du}{dn} + h\,u = \varphi, \tag{10.3}$$

$$\Delta u + f = 0. \tag{10.4}$$

Now let $v$ be an arbitrary function, which I assume only continuous, together with a first-order derivative. We shall have

$$\int \left( v\frac{du}{dn} - u\frac{dv}{dn} \right) d\omega = \int (v\Delta u - u\Delta v)d\tau,$$

so that

$$\int v\,f\,d\tau + \int u\Delta v\,d\tau + \int v\varphi\,d\omega = \int u\left( h\,v + \frac{dv}{dn} \right) d\omega. \tag{10.5}$$

Condition (10.5) is thus a consequence of condition (10.3).

Conversely, if condition (10.5) is satisfied for every function $v$, condition (10.3) will be also satisfied, *provided that u and $\frac{du}{dn}$ are finite, well-defined, and continuous functions.*

But it can happen that in some cases, we are unaware that $\frac{du}{dn}$ is a well-defined and continuous function; we cannot assert then that condition (10.5) entails condition (10.3), and it is even possible that condition (10.3) is meaningless.

Poincaré named condition (10.5) a *modified condition* and asserted (p. 121):

It is obviously equivalent to condition (10.3) from the physical point of view.

This Mémoire of Poincaré contains (p. 70) the first example of an integral inequality between a function and its derivatives:

Let $V$ be an arbitrary function of $x$, $y$, $z$; define:

$$A = \int V^2 d\tau, \quad B = \int \left[ \left(\frac{dV}{dx}\right)^2 + \left(\frac{dV}{dy}\right)^2 + \left(\frac{dV}{dz}\right)^2 \right] d\tau.$$

I will write to shorten:

$$B = \int \sum \left(\frac{dV}{dx}\right)^2 d\tau.$$

I assume first that $V$ satisfies the condition:

$$\int V \, d\tau = 0$$

and I intend to estimate the lower limit of the quotient $\frac{B}{A}$.

The maximum principle is stated on p. 92. Poincaré's principle appears in [59] for the formal construction of the eigenvalues and eigenfunctions of the Laplacian. In [60], Poincaré proved the existence of eigenvalues (for Dirichlet's boundary conditions) using the theory of meromorphic functions (see [50]).

Let us recall that we denote by $L(u)$ the length of the graph of the continuous function $u : [0, 1] \rightarrow \mathbb{R}$. Following Jordan, $L(u) < \infty$ if and only if $u$ is of bounded variation. It follows then from a theorem due to Lebesgue that $u$ is almost everywhere differentiable on $[0, 1]$. In [82], Tonelli proved a theorem equivalent to

$$L(u) = \int_0^1 \sqrt{1 + (u'(x))^2} \, dx \iff u \in W^{1,1}(]0, 1[).$$

A counterexample due to Schwarz, published in 1882 in the *Cours d'Analyse* of Hermite, shows that it is not possible to extend the definition of length due to Jordan to surfaces. Let $z = u(x, y)$ be a nonparametric surface, with $u$ continuous on $[0, 1] \times [0, 1]$. Let $\Omega =]0, 1[\times]0, 1[$ and define, on $X = C(\overline{\Omega})$, the distance

$$d(u, v) = \max\{|u(x, y) - v(x, y)| : (x, y) \in \overline{\Omega}\}.$$

The space of *quasilinear functions* on $\overline{\Omega}$ is defined by

$$Y = \{u \in X : \text{there exists a triangulation } \tau \text{ of } \Omega$$
$$\text{such that, for every } T \in \tau, u\big|_T \text{ is affine}\}.$$

The graph of $u \in Y$ consists of triangles. The sum of the areas of those triangles is called the *elementary* area of the graph of $u$ and is denoted by $B(u)$.

*Lebesgue's area* of the graph of $u$ is defined by

$$A(u) = \inf\left\{\lim_{n \to \infty} B(u_n) : (u_n) \subset Y \text{ and } d(u_n, u) \to 0, \quad n \to \infty\right\}. \tag{$*$}$$

In [83] (see also [53]), Tonelli stated two theorems equivalent to

$$A(u) < \infty \iff \|Du\|_\Omega < \infty,$$

$$A(u) = \int_\Omega \sqrt{1 + \left(\frac{\partial u}{\partial x}\right)^2 + \left(\frac{\partial u}{\partial y}\right)^2} \, dx \, dy \iff u \in W^{1,1}(\Omega).$$

Lebesgue's area is a lower semicontinuous function on $X$. It extends the elementary area: for every $u \in Y$, $A(u) = B(u)$.

In [25], Fréchet observed that Lebesgue's definition allows one to extend lower semicontinuous functions. Let $Y$ be a dense subset of a metric space $X$, and let $B : Y \to [0, +\infty]$ be an l.s.c. function. The function $A$ defined by $(*)$ is an l.s.c. extension of $B$ on $X$ such that for every l.s.c. extension $C$ of $B$ on $X$ and for every $u \in X, C(u) \leq A(u)$.

In [48], Leray defined the weak derivatives of $L^2$ functions and called them *quasi-dérivées*.

In [75], announced in [74] and translated in [78], Sobolev defined the distributions of finite order on $\mathbb{R}^N$, which he called *fonctionnelles*. (A distribution $f$ on $\mathbb{R}^N$ is of *order $k$* if for every sequence $(u_n) \subset \mathcal{D}(\mathbb{R}^N)$ such that the supports of $u_n$ are contained in some compact set and such that $\sup_{|\alpha| \leq k} \|\partial^\alpha u_n\|_\infty \to 0, n \to \infty$, we have $\langle f, u_n \rangle \to 0, n \to \infty$.) Sobolev defined the derivative of a *fonctionnelle* by duality and associated a *fonctionnelle* with every locally integrable function on $\mathbb{R}^N$.

Without reference to his theory of *fonctionnelles*, Sobolev defined in [77] the weak derivatives of integrable functions. Regularization by convolution is due to Leray for $L^2$ functions (see [48]) and to Sobolev for $L^p$ functions (see [77]).

In [69], Laurent Schwartz defined general distributions. In [70], he defined the tempered distributions and their Fourier transform. The treatise [71] is a masterful exposition of distribution theory.

Let $g : \mathbb{R} \to \mathbb{R}$ be a function of bounded variation on every bounded interval. The formula of integration by parts shows that for every $u \in \mathcal{D}(\mathbb{R})$,

$$\int_\mathbb{R} u \, d \, g = - \int_\mathbb{R} u' g \, dx.$$

Stieltjes's integral with respect to $g$ is nothing but the derivative of $g$ in the sense of distributions! Riesz's representation theorem asserts that every continuous linear functional on $C([0, 1])$ is the derivative in the sense of distributions of a function of bounded variation.

## 10.4  Comments

Some general historical references are [15, 19, 29]. We recommend also [46] on Jordan, [52] on Hadamard, [81] on Fréchet, and [38] on Banach.

# References

1. Adams, R., Fournier, J.: Sobolev Spaces, 2nd edn. Elsevier, Oxford (2003)
2. Alberti, G.: Some remarks about a notion of rearrangement. Ann. Scuola Norm. Sup. Pisa Cl. Sci. **29**, 457–472 (2000)
3. Baernstein, A., II, Taylor, B.A.: Spherical rearrangements, subharmonic functions, and ∗-functions in $n$-space. Duke Math. J. **43**, 245–268 (1976)
4. Baker, J.A.: Integration over spheres and the divergence theorem for balls. Am. Math. Mon. **104**, 36–47 (1997)
5. Banach, S.: Sur les opérations dans les ensembles abstraits et leur application aux équations intégrales. Fund. Math. **3**, 133–181 (1922)
6. Banach, S.: Théorie des opérations linéaires. Monografje matematyczne, Varsovie (1932)
7. Bradley, R., Sandifer, E.: Cauchy's Cours d'analyse. An Annotated Translation. Springer, New York (2009)
8. Brezis, H.: Analyse fonctionnelle, théorie et applications. Masson, Paris (1983)
9. Brezis, H., Browder, F.: Partial differential equations in the 20th century. Adv. Math. **135**, 76–144 (1998)
10. Brezis, H., Lieb, E.: A relation between pointwise convergence of functions and convergence of functionals. Proc. Am. Math. Soc. **88**, 486–490 (1983)
11. Choquet, G.: Theory of capacities. Annales de l'Institut Fourier **5**, 131–295 (1953)
12. Daniell, P.: A general form of integral. Ann. Math. **19**, 279–294 (1918)
13. Degiovanni, M., Magrone, P.: Linking solutions for quasilinear equations at critical growth involving the "1-Laplace" operator. Calc. Var. Part. Differ. Equ. **36**, 591–609 (2009)
14. De Giorgi, E.: Definizione ed espressione analitica del perimetro di un insieme. Atti Accad. Naz. Lincei Rend. Cl. Sci. Fis. Mat. Natur. (8) **14**, 390–393 (1953)
15. De Giorgi, E.: Riflessioni su matematica e sapienza. Accademia Pontaniana, Naples (1996)
16. De Giorgi, E.: Semicontinuity Theorems in the Calculus of Variations. Accademia Pontaniana, Naples (2008)
17. de la Vallée Poussin, C.: Sur l'intégrale de Lebesgue. Trans. Am. Math. Soc. **16**, 435–501 (1915)
18. Deny, J., Lions, J.L.: Les espaces du type de Beppo Levi. Annales de l'Institut Fourier **5**, 305–370 (1954)
19. Dugac, P.: Histoire de l'analyse. Vuibert, Paris (2003)
20. Ekeland, I.: On the variational principle. J. Math. Anal. Appl. **47**, 324–353 (1974)
21. Ekeland, I.: Nonconvex minimization problems. Bull. Am. Math. Soc. **1**, 443–474 (1979)
22. Favard, J.: Cours d'analyse de l'Ecole Polytechnique, tome, vol. I. Gauthier-Villars, Paris (1960)

© Springer Nature Switzerland AG 2022
M. Willem, *Functional Analysis*, Cornerstones,
https://doi.org/10.1007/978-3-031-09149-0

23. Fréchet, M.: Sur quelques points du Calcul Fonctionnel. Rend. Circ. Mat. Palermo **22**, 1–74 (1906)
24. Fréchet, M.: Sur l'intégrale d'une fonctionnelle étendue à un ensemble abstrait. Bull. Soc. Math. de France **43**, 248–265 (1915)
25. Fréchet, M.: Sur le prolongement des fonctions semi-continues et sur l'aire des surfaces courbes. Fund. Math. **7**, 210–224 (1925)
26. Gagliardo, E.: Caratterizzazioni delle tracce sulla frontiera relative ad alcune classi di funzioni in $n$ variabili. Rend. Sem. Mat. Univ. Padova **27**, 284–305 (1957)
27. Gagliardo, E.: Proprietà di alcune classi di funzioni in più variabili. Ric. Mat. **7**, 102–137 (1958)
28. Garnir, H.G., De Wilde, M., Schmets, J.: Analyse Fonctionnelle. Birkhäuser, Basel, vol. I (1968), vol. II (1972), vol. III (1973)
29. Giusti, E.: Ipotesi sulla natura degli oggetti matematici. Bollati-Boringhieri, Torino (2000)
30. Golse, F., Laszlo, Y., Viterbo, C.: Analyse Réelle et Complexe. Ecole Polytechnique, Palaiseau (2010)
31. Guiraldenq, P.: Émile Borel, 1871–1956. L'espace et le temps d'une vie sur deux siècles. Librairie Albert Blanchard, Paris (1999)
32. Hadamard, J.: Sur les opérations fonctionnelles. C. R. Acad. Sci. Paris **136**, 351–354 (1903)
33. Hajłasz, P.: Note on Meyers–Serrin's theorem. Exposition. Math. **11**, 377–379 (1993)
34. Hanner, O.: On the uniform convexity of $L^p$ and $\ell^p$. Arkiv för Mathematik **3**, 239–244 (1955)
35. James, R.C.: Orthogonality and linear functionals in normed linear spaces. Trans. Am. Math. Soc. **61**, 265–292 (1947)
36. Jensen, J.L.: Sur les fonctions convexes et les inégalités entre les valeurs moyennes. Acta Math. **30**, 175–193 (1905)
37. Jordan, C.: Sur la série de Fourier. C. R. Acad. Sci. Paris **92**, 228–230 (1881)
38. Kaluza, R., Kostant A., Woycyznski, W.: The Life of Stefan Banach. Birkhäuser, Boston (1996)
39. Kahane, J.P.: Naissance et postérité de l'intégrale de Lebesgue. Gazette des Mathématiciens **89**, 5–20 (2001)
40. Krantz, S.G., Parks, H.R.: The Geometry of Domains in Space. Birkhäuser, Boston (1999)
41. Krantz, S.G., Parks, H.R.: The Implicit Function Theorem. History, Theory and Applications. Birkhäuser, Boston (2002)
42. Lebesgue, H.: Sur une généralisation de l'intégrale définie. C. R. Acad. Sci. Paris **132**, 1025–1028 (1901)
43. Lebesgue, H.: Intégrale, longueur, aire. Ann. Mat. Pura Appl. **7**, 231–359 (1902)
44. Lebesgue, H.: Remarques sur la définition de l'intégrable. Bull. Sci. Math. **29**, 272–275 (1905)
45. Lebesgue, H.: Leçons sur l'intégration et la recherche des fonctions primitives, 2nd edn. Gauthier-Villars, Paris (1928)
46. Lebesgue, H.: Notices d'histoire des mathématiques. L'Enseignement Mathématique, Genève (1958)
47. Lebesgue, H.: Measure and the Integral. Holden-Day, San Francisco (1966)
48. Leray, J.: Sur le mouvement d'un liquide visqueux emplissant l'espace. Acta Math. **63**, 193–248 (1934)
49. Mawhin, J.: Analyse, fondements techniques, évolution, 2nd edn. De Boeck, Paris-Bruxelles (1997)
50. Mawhin, J.: Henri Poincaré et les équations aux dérivées partielles de la physique mathématique. In: L'héritage scientifique de Poincaré, pp. 278–301. Belin, Paris (2006)
51. Maz'ya, V.: Sobolev Spaces with Applications to Elliptic Partial Differential Equations. Springer, Berlin (2011)
52. Maz'ya, V., Shaposhnikova, T.: Jacques Hadamard, a Universal Mathematician. AMS, Providence (1998)
53. Meyer, Y.: Comment mesurer les surfaces? Gaz. Math. **109**, 23–36 (2006)
54. Milnor, J.: Topology from the Differentiable Viewpoint. The University Press of Virginia, Charlottesville (1965)

55. Naumann, J.: Remarks on the Prehistory of Sobolev Spaces, prépublication (2002)
56. Nirenberg, L.: On elliptic partial differential equations. Ann. Scuola Norm. Sup. Pisa Cl. Sci. **13**, 116–162 (1959)
57. Pier, J.P.: Histoire de l'intégration. Masson, Paris (1996); Mathématiques entre savoir et connaissance. Vuibert, Paris (2006)
58. Pietsch, A.: Ein elementarer Beweis des Darstellungssatzes für Distributionen. Math. Nachr. **22**, 47–50 (1960)
59. Poincaré, H.: Sur les équations aux dérivées partielles de la physique mathématique. Am. J. Math. **12**, 211–294 (1890)
60. Poincaré, H.: Sur les équations de la physique mathématique. Rendiconti del Circolo Matematico di Palermo **8**, 57–155 (1894)
61. Riesz, F.: Sur les opérations fonctionnelles linéaires. C. R. Acad. Sci. Paris **149**, 974–977 (1909)
62. Riesz, F., Nagy, B.S.: Leçons d'analyse fonctionnelle, 3rd edn. Gauthier-Villars, Paris (1955)
63. Riesz, M.: Sur les ensembles compacts de fonctions sommables. Acta Szeged Sect. Math. **6**, 136–142 (1933)
64. Roselli, P., Willem, M.: A convexity inequality. Am. Math. Mon. **109**, 64–70 (2002)
65. Roselli, P., Willem, M.: The Lebesgue integral immediately after calculus. Travaux Mathématiques **13**, 61–70 (2002)
66. Royden, H.: Aspects of constructive analysis. Contemp. Math. **39**, 57–64 (1983)
67. Saks, S.: Theory of the Integral. Warsaw–Lvov (1937)
68. Sard, A.: The measure of the critical values of differentiable maps. Bull. Am. Math. Soc. **48**, 883–890 (1942)
69. Schwartz, L.: Généralisation de la notion de fonction, de dérivation, de transformation de Fourier et applications mathématiques et physiques. Annales de l'Univ. de Grenoble **21**, 57–74 (1945)
70. Schwartz, L.: Théorie des distributions et transformation de Fourier. Annales de l'Univ. de Grenoble **23**, 7–24 (1947)
71. Schwartz, L.: Théorie des distributions. Hermann, Paris (1966)
72. Sierpinski, W.: Leçons sur les nombres transfinis. Gauthier-Villars, Paris (1928)
73. Smets, D., Willem, M.: Partial symmetry and asymptotic behavior for some elliptic variational problems. Calc. Var. Part. Differ. Equ. **18**, 57–75 (2003)
74. Sobolev, S.L.: Le problème de Cauchy dans l'espace des fonctionnelles. Comptes Rendus de l'Académie des Sciences de l'U.R.S.S. **3**, 291–294 (1935)
75. Sobolev, S.L.: Méthode nouvelle à résoudre le problème de Cauchy pour les équations linéaires hyperboliques normales. Recueil Mathématique **43**, 39–71 (1936)
76. Sobolev, S.L.: Sur quelques évaluations concernant les familles de fonctions ayant des dérivées à carré intégrable. Comptes Rendus de l'Académie des Sciences de l'U.R.S.S. **I**(X), 279–282 (1936)
77. Sobolev, S.L.: Sur un théorème de l'analyse fonctionnelle. Comptes Rendus de l'Académie des Sciences de l'U.R.S.S. **XX**, 5–9 (1938)
78. Sobolev, S.L.: Some Applications of Functional Analysis in Mathematical Physics. American Mathematical Society, Providence (1991)
79. Steiner, J.: Einfache Beweise der isoperimetrischen Hauptsätze. J. Reine Angew. Math. **18**, 281–296 (1838)
80. Stieltjes, T.: Recherches sur les fractions continues. Ann. Fac. Sci. Toulouse **8**, 1–122 (1894)
81. Taylor, A.E.: A study of Maurice Fréchet. Arch. Hist. Exact Sci. **27**, 233–235 (1982); **34**, 279–280 (1985); **37**, 25–76 (1987)
82. Tonelli, L.: Sulla rettificazione delle curve. Atti R. Accad. delle Sci. di Torino **63**, 783–800 (1908)
83. Tonelli, L.: Sur la quadrature des surfaces. C. R. Acad. Sci. Paris **182**, 1198–1200 (1926)

84. Van Schaftingen, J.: Explicit approximation of the symmetric rearrangement by polarizations. Arch. Math. **93**, 181–190 (2009)
85. Van Schaftingen, J., Willem, M.: Symmetry of solutions of semilinear elliptic problems. J. Eur. Math. Soc. **10**, 439–456 (2008)
86. Vitali, G.: Sul problema della misura dei gruppi di punti di una retta. Bologna (1905)
87. Wolontis, V.: Properties of conformal invariants. Am. J. Math. **74**, 587–606 (1952)

# Index of Notation

© Springer Nature Switzerland AG 2022

M. Willem, *Functional Analysis*, Cornerstones,

https://doi.org/10.1007/978-3-031-09149-0

*Fundamental Theorem of Calculus*
Let $u \in C([a, b])$. For all $a \leq x \leq b$, we have

$$\frac{d}{dx} \int_a^x u(t)dt = u(x).$$

Let $u \in C^1([a, b])$. For all $a \leq x \leq b$, we have

$$\int_a^x \frac{du}{dt}(t)dt = u(x) - u(a).$$

# Index

© Springer Nature Switzerland AG 2022
M. Willem, *Functional Analysis*, Cornerstones,
https://doi.org/10.1007/978-3-031-09149-0

Printed in the United States
by Baker & Taylor Publisher Services